INDIAN BIODIVERSITY FOR THE PRESENT MILLENNIUM:
GLOBAL PROSPECTS AND PERSPECTIVES

Biodiversity - Known and Under-exploited
Threatened Biodiversity - Criteria and Indicators
Assessment and Evaluation Status, Forest Ecosystem Diversity,
Medicinal Flora, Domestic Animal and Marine Biological Diversity
Integration of Traditional Knowledge and Grassroot Innovations
&
Biodiversity Impact Assessment

By

Sas. Biswas

SATISH SERIAL PUBLISHING HOUSE

403, Express Tower, Commercial Complex, Azadpur, Delhi - 110033 (INDIA)
Phone : 011-27672852 Fax : 91-11-27672046
E-mail : ssp@ndf.vsnl.net.in, hkjain1975@yahoo.com
Website : www.satishserial.com

Published by

SATISH SERIAL PUBLISHING HOUSE
403, Express Tower, Commercial Complex,
Azadpur, Delhi - 110033 (INDIA)
Phone : 011-27672852 Fax : 91-11-27672046
E-mail : ssp@ndf.vsnl.net.in, hkjain1975@yahoo.com

First Print 2006

ISBN : 81-89304-09-7

Printed at:
SALASAR IMAGING SYSTEMS, DELHI-35

PREFACE

World can be divided into biodiversity endowed developing nations, developed countries with lesser rich biodiversity but with scientific and technological advances and tools to manage the resources, and the nations with developing economies and presently least concerned with biological resources and their sustainable management. Indian region is one of the fastest growing countries among the developing nations in context to the biodiversity monitoring, inventorization, characterization and sustainable management through integrated approach and concerted efforts.

The biodiversity and genetic resources richness of a country like India is principally on account of diversity in its topography, bioclimatic conditions, biogeography, culture, ethnic community, phytogeographical and zoogeographical entities, socio-economic profiles, and biodiversity- based indigenous traditional knowledge (ITK). The selection system adopted since time immemorial on domestication of wild gene of fauna for animal husbandry and cultivation of land races for agriculture has played a vital role in shaping the present biological diversity and genetic resources of the c ountry. Introduction, invasion and naturalization of species from different parts of the globe have established and developed their regime in various ecological niches. With these significant features in mind, the present treatise has been written.

Contribution is made in the form of chapters dealing with specific aspect of the subject, for example, on the endangered flora and fauna, under-exploited potential of forest biodiversity for the present millennium; threatened forest flora; criteria and indicators for threatened species and habitat; assessment and evaluation status of threatened flora of some well-known countries and case studies with examples; floristic diversity with reference to medicinal plants, and integration of grass root innovation of traditional knowledge for developing conservation strategies. A comprehensive account is in a chapter devoted to the domesticated animal diversity with reference to breeds and races of important animals. The diversity of forest ecosystem is dealt from the point of view of classical works on the forest types' classification, marine biological diversity of the region is dealt in global perspectives. A chapter deals on the biodiversity impact assessment with the prime objective of certain international conventions and a checklist for the assessment process is incorporated following an integrated approach. Bibliographic consultation on each topic is provided with references cited towards the end of each chapter. Glossary of important terms to subjective aspects and common names are provided.

The present treatise has been written, considering the local, regional and global concerns keeping in view the prospects and perspectives, changing perception and integration of different approaches towards the strategic planning, management and the sustainable management of biological diversity. Lacunae in the knowledge are bound to appear as this book is used and to fill these, there is need to interpret, integrate and incorporate more findings from fundamental to applied levels taking into consideration on latest developments on the subject at national and international levels.

Author acknowledges his wife, Rina and son Rishov for their motivation and encouragement, Mr BD Badola for his relentless help in typing the script and to several friends and elders for filling him with inspiration while writing this treatise. To his pet, Sandy, who remained awake with him late in nights during writing this book, author expresses deep love.

Author

Contents

List of Photo Plates

Facing Page No.

Photo: Sas. Biswas, Debojyoti Chakrborty and Anonymous

Chapter I

Biodiversity of Indian Region - Endangered Flora and Fauna

The term 'biodiversity' was first evidently coined in 1985 by Walter, G. Rosen for a meeting, 'The National Forum on Biodiversity' convened in Washington, USA in September 1986. The word 'biodiversity' first appears in the *Biological Abstracts* 'Biosis' database in 1988 with four references, and by the end of April 1994 that number reached to 888 (Harper and Hawksworth, 1996). Diversity characterizes most of the life forms. Biodiversity covers variety of flora and fauna, their ecological roles and the genetic variability they contain. Although natural extinction is part of the overall evolutionary tendencies, the present wave of extinction is manmade. Nearly one million out of 5-10 million species of this planet are thought to be eliminated due to heavy biotic pressure. According to World Conservation Monitoring Centre (WCMC), (Hanski *et al.* 1995) 654 plant and 484 animal species mostly flowering plants and animals such as molluscs 191, birds 115, mammals 58 and others 120 are recorded as having become extinct during the last four centuries. With respect to the species of tropical part of the world the rate of extinction may be higher. Humid and tropical regions particularly of developing countries are highly vulnerable. Forests of these regions are reservoirs of the biodiversity. Infact, biodiversity is fundamental to human life. It is the basic feature of the way in which life forms are structured. With the loss of species and genetic diversity, ecosystems are impaired and the ability to sustain human life is damaged. The floristic and faunistic diversity of India is enormously distributed in the region such as Himalaya, Gangetic plains, desert and Peninsular plateau flanked on both sides by eastern and western Ghats. The richness of the biodiversity of India is largely due to its flora, which covers different ecosystems.

There are as many as 12 biogeographic and 9 botanical regions. Indian region is considered as one of the richest in tropical biodiversity. The country stands as one of the first five countries for the species of plants (45000). The revised and recorded number of flowering plant species number to 17500. For number of species of plants Brazil has 55000, Columbia 45000, China 27000, Mexico 25000, Australia 23000, S. Africa 21000, Indonesia 20000, Peru 20000 and Commonwealth Independent States (erstwhile USSR) 2000. Over 5000 Indian plant species are endemic. The biological diversity of the flora of India is due to an admixture of Indo-Malayan, European, Sino-Japanese, Mediterranean and African elements (Biswas and Dayal, 1994). These elements through intermixing have produced enormous variability; also through the processes of mutation, recombination and subconscious/conscious selection (Khoshoo, 1986). Different bioclimatic regions have provided specific ecological niche to the diversity of the indigenous flora. Plant diversity of the country is represented by species of 17500 flowering plants, Algae 2500, Bryophytes 2843, Gymnosperms 54, Pteridophytes 1022, Lichens 1940, Fungi 23000. Indian region has a rich diversity of many agricultural crops such as cereals, legumes, oil seeds, minor millets, cucurbits, tuberous root crops and banana. The cultivar diversity is so unique that the country is called "*Hindustan centre of origin of cultivated plants*" (Vavilov, 1951).

Fauna

Indian region alone has about 75,000 species of animals out of 132,000 world over , 340 species of mammals, 1200 kinds of birds, 420 reptiles, 140 amphibians, 2000 fishes, 50,000 insects out of 10,85,000 world over, 4000 molluscs (World over 70,000) and other invertebrates distributed in the land mass of 329 million hectares and a coastline of 7516 kms, with Himalayan ranges and innumerable hill tops, plateau, rivers, lakes, deserts, occeans, islands and wetlands (Biswas 1995a). Groombridge (1992) reported threatened diversity of plant and animal in India as 1336 plants, 39 mammals, 72 birds, 17 reptiles, 3 amphibians and 2 fishes, while the rare and threatened species and their geographic races may be much higher than reported in past.

India is the seventh largest country in the world covering an area of 3,287 square km lying entirely in the northern hemisphere. The mainland extends between latitude 8 degree 04 and 37 degree 06N and longitude 68 degree 07 and 97 degree 25 E and measures 3,200 km from north to south between The extreme longitudes. The diversity of life forms of India is distributed in following well-defined regions of the mainland.

The Himalayas occupy the northern part of the country and extends over a distance of 2400 km with a varying depth of 240 to 320 km., the northern plains, approximately 2400 km long formed by the basins of main rivers such as Ganga, Brahmaputra and Indus.

The desert occupies the mid-western part of the country; there is no surface water and rainfall is very scanty.

The peninsular plateau is marked off from the northern plains by hill ranges varying from 450 to 1200 m in height. The peninsula is flanked on one side by the eastern Ghats (600 m) and on the other side by the western Ghats (900-1200 m, and rising at places to over 2450 m). Between the western Ghats and the Arabian Sea lies a narrow coastal strip; the strip is broader between the eastern Ghats and the Bay of Bengal. Nilgiri Hills form the southern part of the plateau and the meeting point of eastern and western Ghats.

The Indian sub-continent presents the most varied physiographic and climatic conditions. The adjoining countries such as Pakistan, Nepal, Bhutan, Bangladesh and Myanmar have common phyto/zoogeographical entities. The floristic and faunistic richness of the country is due to ecological diversity embracing almost all ecosystems from the cold desert vegetation of Tibetan plateau; the alpine meadows of the upper Himalaya; the tropical rain forests of north eastern India and western Ghats, to the xerophytic vegetation of the Thar desert. As many as 12 natural realms are met with in the country.

Distribution of Flora and Fauna - a - Dynamic Phenomenon

The distribution of flora and fauna is a dynamic phenomenon and embraces the whole history of movements of species including both gain and loss of the entire biogeographical area.

The region of northern India is densely populated and under continuous and intensive cultivation for over 4-5000 years. The plains of Uttar Pradesh and Bihar were at one time densely forested and even as late as the 16th century, the Moghuls hunted lions, elephants, buffaloes, bisons, rhinoceros, tigers for game and past time. It is the man who has systematically and most ruthlessly hunted for food, trophy, fur etc. with no regard for conservation. There is a large reduction in the distribution range of floristic and faunistic diversity. The destruction has reached a high magnitude within historic times. The habitats have been destroyed. It is authoritatively recorded (Mani, 1974) that during the past 200 years at least 2 species of birds and 15 spp. of mammals have been practically exterminated. Many of the species have been reduced to the level of geographical relicts rather gradually. For examples, *Gavialis gangeticus* (Gharial) formerly ranged from the river Indus across the whole of the Gangetic Plains of north India to rivers Brahamaputra, Kaladan (Myanmar) and Mahanadi in the Peninsula is at present found sparsely and in isolated and inaccessible places in the river Ganga and its tributary in Nepal - Terai, Bihar and West Bengal. *Crocodilus palustris* (Indian crocodile) - an inhabitant of marshy areas and small pools is eroded of its natural habitat. The reptiles have been mercilessly hunted in India for their skin for foreign markets, manufacture of shoes, belts, bags and fancy products. Indian crocodile has almost been exterminated from its natural habitat but for its some religious significance the species has been saved in some parts, e.g., Puskar Lake (Ajmer), besides preservation in Crocodile Banks in different parts of India. In past there was a fashion to keep crocodiles around fortresses etc. as means of defense, but with latest warfare technology there has been onslaught on these species. Due to the destruction of sal forests and swampy areas being reclaimed for agriculture the population of the swamp deer has reduced drastically. In Madhya Pradesh the species is protected in Kanha National Park. Similarly many plant species of ornamental and economic value such as *Dioscorea deltoidea* from NW Himalaya, blue vanda, *Calanthe whiteana, Aerides ringens, Cyathea* spp., lady slipper orchids, *Coptis teeta*, *Hedychium greeni*, *Coelogyne albo-lutea*, *C. assamica*, *C. treutleri* etc. from NE India and *Paphiopedilum druryi*, *Ceropegia* spp., *Pavetta* spp., *Syzygium* spp. etc. from South India have become threatened with extinction due to severe biotic influence.

Distributional Patterns of Flora and Fauna

The distribution of fauna of India has been greatly influenced by the recent origin of transit conditions such as Himalaya, stable and senile Peninsula, biogeographially important Indo-Gangetic Plain found gateway. The latter has only the rapidly vanishing species from the south and east. The characters of fauna today hardly give any clue to its composition, patterns of distribution and diversity just what it was 5000 years ago. The distribution patterns of different species do not correlate with the physical features, physiography and rainfall.

Due to the onslaught on the original flora and fauna there is a most striking or lack of endemic elements. The fauna of the plains was in reality largely a spill over from that of the Peninsular region. The eastern parts of the plains are largely dominated by derivatives of the fauna from the South Chinese and Indo-Chinese and Malayan areas. The plains of western part are increasingly dominated by the derivatives of the Mediterranean-Ethiopian elements. The discontinuities in the fresh-water species of the rivers Indus and Ganga are observed quite significantly. The fresh-water vertebrates of the rivers-mentioned are identical in many cases; some of these occur also in the rivers of the Peninsula (Godavari, Mahanadi). Similar is the feature for a number of species of the rivers Ganga, Brahmputra, Indus and their tributaries. It is believed that the area where we have now the Indo-Gangetic plains was formerly a narrowing channel of sea. The streams carried down the alluvium or silt from the new mountains to fill the channel.

The disjunct distribution of Indo-Chinese and Malayan flora and fauna in the Peninsular south was well known to the early naturalists. The elements typical of Himalayan region are found on other hill tops of India (Biswas, 1995). The disjunct distribution was correlated more with the conditions of the atmospheric humidity and precipitation than to temperature. The present day discontinuity is essentially a relict of former continuous distribution. The disjunct elements found in Himalayan and North eastern, and South Indian regions such as *Michelia champaca, Rhododendron a rboreum* (form *nilagirica*), *Tetrameles mudiflora, Lagerstroemia regina* (Syn. *L. speciosa*), *Prinsepia utilis*, etc. are some striking examples. The extensive and continuous ranges of a number of humid tropical forest forms have broken into isolated tracts due to topographical/climatic changes/gradual retraction and regression. In some cases man has annihilated and thus resulted in the present day discontinuity. In certain cases the influx of the outside fauna and the rapid colonization eventually resulted in a rather pronounced discontinuity.

Forest Types - Lifeline of Flora and Fauna

Bioclimatically the plants (flora) of India are distributed in (i) Tropical, (ii) Temperate and (iii) Alpine types. From the view point of general vegetation and floristic elements, country can be broadly divided into following botanical regions:

1. North Western Himalayas (Oak, Pine, Deodar, Fir, Junipers, Sal and other broad leaved spp. in lower hills).

2. Eastern Himalaya (Dipterocarps, Laurels, Myrobalans, Oaks, *Machilus*, *Persea*, Conifers in higher reaches).

3. Western Arid region (Scattered trees and bushy and thorny plants e.g. *Calligonum, Capparis, Prosopis, Acacia, Euphorbia, Ziziphus, Cassia*).

4. Gangetic Plains (Sal, Haldu, *Butea, Mallotus*, Jamun, fodder yielding grasses).

5. Eastern India (species of evergreen forests; Khasi pine; bamboo, Orchids etc.).

6. Deccan plateau (Tropical thorn forests, dry forests with *Pterocarpus, Adina, Diospyros, Aegle*, Dhak, Red sander, Teak, Semul, Sal, Mahua etc.).

7. Western Ghats:

   (i) Moist deciduous forest - *Terminalia, Dalbergia, Pterocarpus* etc.

   (ii) Tropical evergreen forests - *Hopea*, Kadam, *Xylia, Terminalia,Lagerstroemia* etc.

   (iii) Tropical wet evergreen forests - *Mesua, Cullenia, Dipterocarpus, Palaquium, Canarium, Garcinia, Cinnamomum* species.

   (iv) Subtropical broad-leaved hill forests of Nilgiris - *Syzygium, Elaeocarpus, Celtis* etc.

8. Eastern Ghats - Tropical dry deciduous forests of thorn forests. Evergreen forests have *Trema, Vitex, Diospyros, Syzygium, Wendlandia* etc.

9. Andaman & Nicobar Islands - *Dipterocarpus, Pterocarpus* (Padauk), *Terminalia bialata, Canarium, Albizia, Chukrassia*.

10. Littoral Forests - The coastal areas have mangrove vegetation - *Rhizophora, Heritiera, Avicennia, Sonneratia, Pandanus, Thespesia, Nypa* etc.

Threatened Biodiversity

According to WCMC (World Conservation Monitoring Centre), world over 26,106 species of plants fall under varying degree of threats (Endangered: 3632; Vulnerable: 5687; Rare: 11485; Indeterminate:

PLATE I

Threatened Habitats of Biodiversity

1. Mining and landslides
2. Submergence by Hydro dam
3. Invasion of weeds
4. Rock blasting for roads

5302).Hanski *et al.* (1995) believe that the species most threatened with extinction are: (i) large organisms, (ii) top of food chain as per ecological theory, (iii) small population sizes, (iv) small range or distribution (v) those tied in some aspect of their biology to habitats or ecosystems that are threatened, (vi) evolved in isolation (island fauna and flora), (vii) poor dispersal and colonization abilities (vii) colonial habits, (ix) migratory species, (x) dependent on unreliable resources, and (xi) with little evolutionary experience of disturbances.

Restricted distribution (endemism) of flora and fauna of the country has become the subject of serious concern since such elements (endemics) are highly threatened with extinction. As per a report (*Vide* National Action Plan on Biodiversity, 1997) 807 species of Molluscs, 619 Protozoa (Parasitic: 532 & Free living 87), 152 Reptilia, 85 Amphibian, 78 Pisces (Freshwater: 64 and Marine 14), 50 Aves (Bird), 37 Mammals and 9 of Lepidoptera; and nearly 5000 species of angiospermic plants are endemic.

Rare and Threatened Flora

IUCN (International Union for Conservation of Nature and Natural Resources) has estimated that about 25,000 plant species are on the verge of extinction world over. The flora of India is in varying degrees of threat. As many as 3000 plant species are threatened with extinction whereas documentary evidence reveals *ca* 1300 taxa. The species such as *Aerides ringens* (Tamil Nadu), *Satyrium nepalense* (Tamil Nadu), *Pholidota articulata* (Uttaranchal), *Coelogyne albolutes, C. assamica, C. treutleri* (Sikkim, Assam), *Madhuca bourdillonii* (SW Ghats), *M. insignis* (W. Peninsula), *Carum villosum* (N. Champaran, Bihar), *Urginea polyphylla* (Deccan Peninsula), *Ilex gardeneriana* (Sisapa Ghat, Tamil Nadu), *Begonia canarana* (Canara), *B. watti* (Nagaland), *Ceropegia arnottiana* (Khasi hills), *C. fantastica* (W. India), *Cyclea wattii* (Nagaland), *Dennstaedtia elwesii* (Sikkim-Lachen), *Deuxia simlensis* (Shimla, Himachal Pradesh), *Dipcadi concanense* (S. India), *D. reidii* (W. Himalaya), *Eragrostis rottleri* (E. coast - Tranquebar), *Eriochrysis rangacharii* (Paikara - Nilgiri), *Eugenia argentea* (Wynaad-Kerala), *Ligusticum alba-alatum* (Chhotanagpur Plateau, Jharkhand), *Ophiorrhiza barnesii* (Travancore), *O. brunonis* (S. India hills), *O. caudata* (Kerala), *O. pykarensis* (Nilgiri), *O. radicans* (Kerala), *Indigofera byansghatensis*

(Garhwal, Uttaranchal), *Pavetta oblanceolata* (S.W. Ghats), *P. wightii* (Tamil Nadu), *Pimpinella evoluta* (Nagaland), *P. pulneyensis* (Pulney hills), *Pleione lagenaria* (Meghalaya), *Psychotria tylophora* (Nicobar Is.), *Sageraea grandiflora* (Konni Forest - Kerala), *Sterculia khasiana* (Khasi hills, Meghalaya), *Syzygium palghatense* (Palghat hills, Tamil Nadu & Kerala), *Urginea polyphylla* (Deccan Peninsula), *Vanda wightii* (Nilgiri), *Zeuxine pulchera* (Sikkim, Meghalaya), *Bambusa copelandi* (NE & NW India - Biswas and Jain, 2002). *Bunium nothum* (Nilgiri) seems to have probably become extinct. Besides, many species of plants of biological value and economic importance have become rare and endangered. For example, Orchids (*Paphiopedilum* spp., *Vanda coerulea* - the blue vanda) *Nepenthes khasiana* (Pitcher plant), Tree Fern (*Cyathea gigantea, Magnolia pterocarpa, Sapria himalayana* (close ally of famous *Rafflesia* of Malaysia), *Rhododendron* spp., *Dischidia rafflesiana* (Botanical curio), *Coptis teeta, Meconopsis betonicifolia* etc. The causes of threats to the flora are destruction of their habitats, over-exploitation, competition from introduced plants etc. (Biswas, 1980; Biswas *et al.* 1996).

Significance Wise some Endangered Indian Plants of **Ornamental, Biological and Economic Value** (Based on Biswas, 1990; Biswas & Dayal, 1994 and Biswas *et al.* 1996; Tewari, 1994).

Medicinal and Aromatic Plants: *Aconitum falconeri, A. ferox, A. deinorrhizum, Acorus calamus, Aerva wightii, Alpinia galanga,Aquilaria malaccensis, Artemisia amygdalina, Asparagus jacquemonti, Atropa acuminata, Chlorophytum borivilianum, C. glaucum, Colchicum luteum,Coleus forskohlii,Commiphora wightii,Coptis teeta, Dioscorea deltoidea, D. prazeri, D. rogersii, D. wightii,Gentiana kurroo, Gloriosa superba, Heliotropium keralense, Indigofera caerulea, Inula racemosa, Iphigenia magnifica, Janakia arayapathra,Kaempferia galanga, Kalanchoe roseus, Lamprachaenium microcephalum, Nardostachys grandiflora, Nilgirianthus ciliatus, Ochneinauclea missionis, Orchis latifolia, Orophea uniflora, Panax pseudoginseng, Pauia belladonna, Pittosporum eriocarpum, Plectranthus bishopianus, Podophyllum* ssp. *Rauvolfia* spp., *Rheum* spp., *Saussurea lappa, Taxus baccata* ssp *wallichiana, Stephania andamanica, Urginea congesta, Vanilla wightiana*. A separate chapter is contributed on the subject in this treatise

Potential Forest Tree Genetic Resource: *Abutilon lawsonii, A. ranadei, Acacia campbellii, Acer caesium, Acer hookeri* var. *majus, A. osmastonii, A. sikkimensis* var. *serrulatum, Actinodaphnae bourneae, Aglaia perviridi, Ailanthus kurzii, Albizia thompsonii, Ammania desertorum, Anogeissus sericea* var. *nummularia, Artabotrys nicobaricus, Buchanania platyneura, Bombax insigne* var. *polystemon, Bridelia kurzii, Buchanania barberi, Canarium mannii, Cryptocarya ferrarsii, Cynometra beddomei, C. bourdillonii, Dipterocarpus kerrii, Drypetes andamanica, Echinocarpus nimmonii, Elaegnus coniferta* ssp. *dendroidea, Elaeocarpus acuminatus, E. prunifolius, E. recurvatus, E. singampattiana, E. venustus, Eriolaena lushingtonii, Eugenia argenta, Ficus andamanica, Gleditsea assamica, Hippocratea nicobarica, Hopea jacobi, Humboldtia unijuga, Hydnocarpus macrocarpa, Hydnocarpus macrocarpa, Ilex embelioides, Ilex gareneriana, Inga cynometroides, Kingiodendron pinnatum, Lagerstroemia hypoleuca, Lagerstroemia minuticarpa, Litsea leiantha, Madhuca boudillonii, M. diplostemon, M. insignis, Magnolia pterocarpa, M. griffithii, Mangifera andamanica, Mesua manii, Meteoromyrtus wynaadensis, Michelia punduana, Miliusa nilagirica, Myrstica andamanica, Ormosia glauca, Orophea uniflora, Palaquium boudillonii, Poeciloneuron pauciflorum, Populus gamblei, Potameia paradoxa, Pterocarpus santalinus, Pterospermum obtusifolium, Pterospermum reticulatum, Rhus hookeri, Rotala ritchiei, Sterculia khasiana, Symplocos odoratissima, Syzygium andamanicum, S. boudillonii, S. courtallense, S. palghatense, S. travancoricum, Uvaria nicobarica, Vateria macrocarpa,*

Bamboos: *Arundinaria clarkei, Bambusa affinis, Bambusa copelandii* (Perhaps extinct), *Bambusa jaintiana, Bambusa kingiana, Bambusa pseudopallida, Chimonobambusa jaunsarensis, Dendrocalamus hookeri* var. *parishii, Dendrocalamus sahnii, Dendrocalamus sericeus, Dinochloa gracilis, Drepanostachyum falcatum* var. *glomerata, Neomicrocalamus mannii, Ochlandra scriptoria, Ochlandra travancorica, O. beddomei, O. talbotii, Pseudoxytenanthera bourdillonii, Pseudoxytenanthera ritcheyi, Schizostachyum beddomei, S. capitatum* var. *decomposita, Semiarundinaria pantlingii* etc.

Rattans and Palms: *Bentinckia condapanna, B. nicobarica, Calamus dilaceratus, C. erectus, C. flagellum, C. floribundus, C. gamblei, C.*

huegelianus, C. inermis, C. khasianus, C. kingianus, C. latifolius, C. leptospadix, C. mishmeensis, C. vattayila, Corypha macropoda, Daemonorops jenkinsianum, Korthalsia rogersii, Livistona jenkinsiana, Phoenix robusta, P. rupicola, P. acaulis, Pinanga andamanensis, P. manii, Plectocomia himalayana, Trachycarpus takil, Wallichia triandra etc.

Edible Plants: *Amorphophallus longistylus, A. oncophyllus, Ceropegia* spp. (30), *Garcinia caddelliana, G. kingii, Oryza indandamanica, Oryza nivara, Vigna khandalensis* etc.

Conifers and Cycads: *Abies delavayi, Amentotaxus assamica, Cephalotaxus sinensis* (cult.), *Cycas beddomei, Gnetum ula, Picea brachytyla, Pinus bhutanica, Podocarpus neriifolium, P. wallichianus, P. gracilior* (cult.) etc.

Spectacular Ornamentals: *Aspidopterys canarensis, A.oxyphylla, Begonia aborensis, B. aliciae, B. anamalayana, B. canarana, B. kashmeriana, B. lambertii, B. lushaiensis, B. osmastonii, B. scutata, B. wattii* and other species, *Berberis affinis, Capparis cinerea, Catamixis baccharoides, Clematis apiculata, C. boudillonii, C. theobromina, Corybas purpureus, Cotoneaster buxifolius, Ensete superba, Eremostachys superba, Impatiens anaimudica, I. johnii, I. macrocarpa, I. munnarensis, I. neobarnesii, I. nilagarica, Inula racemosa, Ipomoea clarkei, Jasminum unifolilatum, Kalanchoe olivacea, K. rosea, Lilium macklinae, Meconopsis betonicifolia, M. latifoia, Olax nana, Orchids, Rhododendron* (43 spp.): *R. arizelum, R. dalhousiae, R. edgeworthii, R. nivale, R. nuttallii, R. santapaui, R. senaulum, Rosa clinophylla, Sophora mollis, Strobilanthus dupenii, Strobilanthus hallbergii, Thalictrum dalzellii,* etc.

Orchids: *Arachnanthe cathcartii, A. clarkei, Arundina graminifolia, Calanthe biloba, C. whiteana, Cymbidium eburneum, C. giganteum, C. grandiflorum, C. lowianum, C. macrorhiza, C. sikkimense, C. tigrinum, Cyperipedium elegans, C. himalaicum, C. tibeticum, Dendrobium densiflorum, Eulophia mackinonii, E. nicobarica, Galeola falconeri, Habenaria andamanica, Malleola andamanica, Paphiopedilum druryi, P. fairrieanum, P. hirsutissimum, P. insigne, P.*

spicerianum, P. venustum, P. villosum, Pleione humilis, Renanthera imshootiana, Vanda coerulea, Vanilla pilifera, V. wightiana, etc.

Ferns: *Angiopteris evecta, Botrychium virginianum, Brainea insignis, Cyathea gigantean, Diplazium travancoricum, Dipteris wallichii, Helminthostachys zeylanica, Ophioglossum pendulum, Osmunda regalis, Platycerium wallichii, Schizaea digitata* etc.

Biological curios: *Aldrovanda vesiculosa, Balanophora dioica, Dischidia benghalensis, Dischidia rafflesiana, Drosera burmani, D. indica, D. peltata, Entada pursaetha, Helwingia himalaica, Magnolia pterocarpa, Mitrastemon yamanotoi, Monotropa uniflora, Nepenthes khasiana, Psilotum complanatum, P. nudum, Sapria himalayana, Thylacospermum rupifragrum,* etc.

The expansion of agriculture, urbanization, industrialization and overgrazing have drastically modified the distribution patterns and richness of the floristic and faunistic diversity of India.

Rare and Threatened Fauna: Globally 5366 species fall under varying degree of threats (mammals: 533; birds: 862; reptiles: 257; amphibians 133; fishes: 934, and invertebrates: 2647) are threatened with extinction. Grimmett *et al.* (1999) mention about the 10% of world's avifauna i.e. 1111 as threatened. It deserves a mention here that among the rare butterflies on nature reserves in Britain between 1960 and 1982, 100 percent of populations of *Lycaena dispar, Maculinea arion* and *Carterocephalus palaemon* went extinct (see Hanski *et al.*, 1995). Habitat alteration, use of pesticides, oil spills in oceanic ecosystem, excessive siltation etc. have made an adverse impact on the population of invertebrates and the dependent species. Indian region represents nearly 75,000 species in its faunistic diversity covering 6.5 per cent of world's fauna. Nearly 62 per cent of animal species of India is endemic.

Major Groups: **Mammals:**

No. of Species threatened/group: 27; species endemic: 37.

No. of Species: 340

Cat Family

Important Species

Tiger (*Panthera tigris*): There were 40 thousand at the turn of 20th century, 1800 in 1973, some 4000 in the early 1980s and 2500-2750 in the 1990s; but due to heavy poaching and sporting (game) the species became threatened with extinction (*vide* Hindustan Times, 13 June, 1997). Indian lion (*Panthera leo*) - Confined to Gir forests of Gujarat now. Leopard (*Panthera pardus*) and Snow leopard (*P. unica*) despite their camouflaging habit are threatened. *Cheetah* (Hunting leopard) became extinct from Indian region by the turn of 20th century. Indian cheetah (*Actinomyx jabatus*) became extinct since it was last noticed in some part of forests of Tamil Nadu in 1952.

Desert cats (*Felis sylvestris*), Lynx (*Felis lynx*), Caracal (*Felis caracal*), Clouded leopard (*Neoelis nebulosa*), Jungle Cat (*Felis chaus*), Rusty spotted cat (*Felis rubiginosa*), Palla's cat (*Felis manual*) and Fishing cat (*Felis marmorata*) are rare and threatened.

Civet Cats: Malabar civet (*Viverra megaspilla*) in coastal areas of Kerala, Tiger civet (*Prionodon pardicolor*) in Eastern Nepal to Arunachal Pradesh. Binturang (*Arctitis binturung*) is found in some limited area in eastern Himalaya.

According to Dr. Peter Jackson, Chairman of the Cat Specialist Group of IUCN the tiger is threatened with extinction as it can be sold for up to 15,000 dollars and there are poachers to earn that amount at any cost. The bones of tiger have high medicinal value in China. One of the main reasons of threat to tiger is due to its position on the top of food chain in a natural ecosystem.

Primates: Slow loris (*Nycticebus coucang*) is found in the dense forests of NE India. Slender loris (*Loris tardigrandus*) - in tropical rain forests of southern India. Lion - tailed macaque (*Macaca silenus*) found in evergreen forests of western Ghats; pig-tailed macaque (*Macaca nemestrina*), stump-tailed macaque (*Macaca arto-toides*) and Assamese macaque (*Macaca assamensis*). *Presbytis geei* (Golden Langur) is distributed in NE India and eastern Himalayas along Bhutan. Nilgiri Langur (*Presbytis johni*) is found in shola forests of Nilgiri (Tamil Nadu). Removal of trees and distorted crown and canopy of forests sometimes result in to death of macaque and langur while attempting to jump across.

Philidota: Chinese pangolin (*Manis pentadactyla*) is found in NE Indian forests and in other parts. Indian pangolin (*M. crassicaudata*) bears broad scales on its body.

Rodents: Phayer's flying squirrel (*Hylopetes phayrei*), Particoloured flying squirrel (*Hylopetes oblongiger*), Smaller Kashmir flying squirrel (*Hylopetes fimbriatus*), Gray's flying squirrel (*Petaaurista nobitis*), Hodgson's flying squirrel (*Petaurista magnificus*), White flying squirrel (*Petaurista alborufus*), Lesser giant flying squirrel (*Petaourista elegans*), Common giant flying squirrel (*Petaourista petaurista*), Travancore flying squirrel (*Pentinomys fuscocapillus*) and Woolly Kashmir flying squirrel (*Epetaurus cinerus*).

Marmots: Long-tailed marmot (*Marmota caudata*) and Himalayan marmot (*Marmota bobak*).

Bears: Himalayan brown bear (*Ursus arctos*), Sloth bear (*Mellursus ursinus*) and Malayan sun bear (*Helarctos malayanus*), last two found in forests of NE region.

Perisso dactyla: One-horned rhinoceros (*Rhinoceros sondiacus* and Asian two-horned rhinoceros (*Didermoceras sumatrensis*) - Extinct.

Rhinoceros (*Rhinoceros unicornis*) - North Bengal, Assam and lower hills of Arunachal Pradesh. During the reign of Babur (1505-1530), the Mogul Emperor, one-horned rhino was distributed across the north and north-east of India to Peshawar (border of Pakistan and Afghanistan). The decline of its population in past was due to sport of killing by European hunters (Kushwaha, 1997). Present population is over 1000 individuals risen from nearly extinct with few left in Kaziranga and Jagdalpara tracts of eastern India in 1904.

Indian Wild Ass (*Asinus hemionus*) - Confined to Rann of Kutch in Gujarat.

Kiang (*Asinus kiang*) - Tibetan wild ass found in the trans - Himalayan cold desert of Ladakh, Lahul and Spiti.

Proboscidia: Indian Elephant (*Elephas maximus*).

Sirenia: Dugong (*Dugong dugong*) - Once thrived along the Gulf of Kutch, Gulf of Mannar, coastal areas of Kerala and the Andaman and Nicobar Islands.

Pigs: Pigmy hog (*Porcula sylvania*) and Nicobar wild pig (*Sus scrofa nicobaricus*),

Canines: Indian wild dog (*Cuion alpinus*), Striped hyaena (*Hyaena hyaena*), Common wolf (*Canis lupus*), Red fox (*Vulpes vulpes*), Indian fox (*Vulpes bengalensis*).

Aquatic mammals: 25 species. Ganga dolphin (*Platanista gangetica*), an endangered species was once common in rivers of northern and eastern India. The species exists with a population between 1800-2000. A 2004 census carried out by WWF - India, NGOs and State Forest Departments, shows that the number has dropped to one-third of what was believed to be the dolphin count in the early 1980's. WWF - India adopted a conservation program in 1997 which showed the number of dolphin doubled from 21 to 42 in 1991 in the Bijor - Narora stretch (*Times of India*, 21.7.2005). The species is distributed in the Ganga - Brahmaputra - Meghna river systems. The major threats to the species are poaching, constuction of large dams, habitat degradation and pollution.

Other mammals

Gaur (*Bos gaurus*) - Foot hill forests of western Ghats and NE India.

Wild Yak (*Bos grunnieus*) - Confined to Ladakh (Chengchemo valley).

Hispid hare (*Carpolagus hispidus*) - Found in terai tract of eastern Himalaya and Brahmaputra valley of Assam.

Urial or Shapu, (*Ovis vignei*), Tibetan sheep (*Ovis amoni*), Himalayan tahr (*Hemitragus jemlahicus*, Marco Polo sheep (*Ovisammon polli*) are some of the rare and spectacular faunistic diversity of the country.

Bharal (*Ovis nahura*), Himalayan ibex (*Carpa ibex*) and Markhor (*Carpa falconeri*).

Antelopes and Deers:

Brown Antlered deer (*Cervus eldi eldi) - Sangai* is mainly confined to the Keibul Lamjao National Park area just south of famous Loktak Lake of Manipur. It is one of the most threatened mammals of the world. As per the census of March 1979, a pitifully 30 *sangai* were counted. What makes Keibul Lamjae unique is the fact that the natural habitat of this deer floats. It is because the dense morass of rotting vegetation and humus (*phumdi*) which float on water.

Barasingha or swamp deer (*Cervus elephus)* - Found in the Dachigam area of Kashmir valley.

Tibetan antelope (*Pantholops hodgsoni*) - Found in the cold desert of Ladakh.

Tibetan gazelle (*Procarpa picticaudata*) - Occurs in higher and trans - Himalayan region.

Indian gazelle or Chinkara (*Gazella dorcas bennatti*) - Found in the arid and semi-arid areas of Rajasthan.

Black buck (*Antelope cervi-carpa*) - Plains of western and central India. Protected by the Bishnoi community in Rajasthan and Haryana.

Himalayan musk deer (*Moschus moschiferus*) - found in higher reaches of Himalaya.

Alpine musk deer (*Moschus sifanicus*) - Found in higher elevation of Himalaya.

Reptiles: 17 reptiles are recorded as threatened among 156 endemic species.

Gharial and Crocodiles: Gharial (*Ghavialis gangeticus*) - Found in many of the large rivers of India. It is the only species of *Gavialidae* in the world. The species has been saved from brink of extinction during the past over two decades. During the last 1970s the species had a ramnant population of 30 individuals in the Girwa river of Bahraich Forest Division of Uttar Pradesh in northern India, whereas over 1000 of gharial have been developed at different breeding centres in the country.

Marsh crocodile (*Crocodylus palustris*) - Distributed in different river systems of the country.

Estuarine crocodile (*Crocodylus porosus*) - Found in the estuaries and deltas of rivers.

Snakes: King cobra (*Naja hannah*) - Royal snake (*Spalerosophis diadema*), Striped keel-back (*Amphiesma stolata*), and Indian egg eating snake.

Lizards: Water monitor, clouded monitor, yellow and desert monitor.

Tortoises, Turtles and Terropins: Smith's Kachua (*Kachuga dhongoka*), Starred tortoise (*Testudo elegans*), Thurgi (*Hardella thurgi*), Ganga soft sheel tortoise (*Trionyx gangeticus*).

Amphibians: Three amphibians are threatened. 110 spp. are endemic to Indian region. Viviparous toad (*Nectophyrne tuberculosa*) - Occurs in coastal forests. Indian salamander, a kind of tailed amphibian is found in NE India and North Bengal Himalaya. Recent discovery of *Nasikabatrachus sahyadrensis* from Western Ghats - a frog with purple colour, tiny eyes and protruding snout belongs to dinosaur era. The species belongs to Seychelles branch of the frog - family Sooglossidae found near Madagascar (*India Today* November 3, 2003).

Pisces: Three species of fish are reported to be threatened. According to Joshi (1999) 5 species out of 34 indigenously found in Kumaon rivers of

Uttaranchal Himalayas are extremely rare and threatened with extinction. The threats are largely due to over fishing, dynamiting, toxicants and heavy silting of rivers.

Groups	Sps.	Groups	Sps.
Pheasants, Monal, Junglefowl, Peafowl.	13	Pigeons, Doves, Sandgrouse	38
Woodpeckers, Barbets, Cuckoos	48	Kites, Hawks, Falcons, Buzzard etc.	34
Hornbills	10	Eagles, Vulture	25
Parrots and Parakeets	12	Shrikes, Thrush, Minivet, Robin/Chats	81
Owls, Nightjars and Frogmouths	43	Drongo, Flycatcher, Tits, Wren	61
Bulbuls, Prinias etc.	40	Warbler, Bablers.	109
Myna	8	Sparrows, Snowfinch	14
Sunbirds, Flowerpeckers, Weaverbird	25		
Munia, Finckes, Buntings	54		

Birds: Out of 1235 species of Indian avifauna as many as 176 species of birds are considered to be endemic. Largely the birds are dependent on forest ecosystem. Some of the most important groups of birds typical of forest areas and dependent on forest ecosystem are mentioned in the above table.

According to Alfred *et al.* (2001), there are 767 water birds (as recognized by Ramsar Convention) world over. Out of 328 Asian water birds, 245 are known from India. Northern India has 180 forms of resident and migratory birds distributed in as many as 37 wetland sites.

Nearly 100 species of Indian birds are threatened. It is estimated that the extinction of a bird causes the loss of 90 species of insects, 35 species of plants and 3 species of fish. Infact in the biological pyramid birds structure the intermediate strata. Egg shell thinning of the birds such as vultures, eagles, adjutant storks etc. has resulted into the decline of their population. The thinning of egg shells has appeared due to medicines used as pain killer in veterinary practices, the traces of which remain with the carcass of the dead animals on which the birds scavenge. The major threats to the survival of avifauna are due to pesticide poisoning, habitat destruction, hunting pressure, predation and competition from alien species.

Mountain quail (*Ophrysia superciliosa*) is reported once to be found in western Himalaya (Mussoorie and Nainital 2-4000 m), it was last observed in 1876. The species is now 'presumed extinct' (Salim Ali, 1964; Grimmett *et al.*, 1999). Pink - headed duck (*Rhodonessa caryophyllacea*) - Last authentic record in June 1935 from Darbhanga in Bihar. Population started dwindling as early as 1878, although until about 1890 a half dozen or so turned up in Kolkata (Calcutta) bird market (Hathibagan) and elsewhere every season. Possibly the bird is already extinct (Salim Ali, 1964). Out of 1111 bird species threatened with extinction as published by Birdlife International in 1994, 78 species typical of Indian subcontinent are mentioned (see Grimmett *et al.* 1999).

White-winged wood duck or *deohans (Cairina scutulata*) - Found in water bodies of NE India.

Andaman teal - Found only in the Andaman group of islands. Among the endangered cranes, Siberian crane (*Grus leucogeranus*) is migrant to the Bharatpur Bird Sanctuary and other wetlands of India. Among the migratory swans, the Whopper, Whistling and Mute swan deserve special mention.

Little Bustard, Social Lapwing, Bengal Floricans and Great Indian Bustard are some of the most threatened avifauna. Nicobar Megapode (*Megapodius freycinet*), Andaman Wood pigeon, Andaman Cuckoo dove, Nicobar Parakeet, Nicobar Pigeon endemic to Nicobar groups of islands are endangered. Nicobar pigeon is reported to have affinity with extinct bird, dodo, as per a study on its DNA profile.

Among the hornbills, Narcondam hornbill (*Aceros narcondami*) found in Andamans, Rufous-necked hornbill (*Aceros nipalensis*) of Himalayan region, Wreathed hornbill of Himalayan foot hills and White-throated hornbill of eastern India are rare and threatened species.

Large falcons, Red-necked falcon, the White-bellied eagle, Golden eagle, Bearded vulture, Snow cock, Koklass pheasant, Mishmi hills monal, Himalayan monal are some of the threatened species of birds needing attention for conservation.

PLATE II

Spectacular Insect Diversity of India

1. *Attacus* – Meghalaya
2. *Papilio*- Uttaranchal
3. Stick Insect- Meghalaya
4. *Cyristus* - Arunachal

Andaman hawk owl, Andaman scoops owl (*Otus balli*); Forest owlet (*Athene blewitii*) was last recorded in 1914.

Some spectacular, rare and threatened butterflies and moths of India

A ban is imposed on the national and international trade on the export of spectacular butterflies. Wildlife (Protection) Act (1972) as ammended in 2001 cites 128 taxa of butterflies and moths under Schedule I of the Act. Some of striking diversity of this group of butterflies and moths needing attention of conservationists are following:

African migrant (*Catopsolia florella*)
Aporia leucodice - Kumaon
Argynnis altissima - NW Himalaya
Argynnis kamala - NWHimalaya (Kumaon)
Atlas moth (*Attacus atlas*)
Blue admiral (*Vanessa canace*) - Himalaya and South India
Blue triangle (*Graphium sarpedon*)
Colour sergeant (*Parathyma nefte*)
Common club tail (*Parides coon*)
Common faun (*Faunis canens*)
Common imperial (*Charitra freja*)
Common map (*Cyrestis thyodames*)
Cyllogenes suradeva - E. Himalaya
Cyllogenes janetae - Assam
Discophora sondaica ssp. *indica* - NE India
Elymnias malelas - Kumaon Himalaya (Uttaranchal)
Elymnias nesaea - Sikkim
Erebia hybrida - Uttaranchal Himalaya
Erebia scanda optim - Sikkim
Erebia vasudeva vasudeva - NE India
Ethipe himachala - NE India
Golden emperor moth (*Loepa katinka*)
Great egg-fly (*Hypolimnas bolina*)
Green dragon tail (*Lamproptera meges*)
Great orange tip (*Hebomoia glaucippe*)
Indian moon moth (*Actias selene*)
Jewelled nawab (*Polyura delphis*)

Kohi-I-Noor Butterfly (*Amathuxidia amythaon*) - Sikkim
Northern jungle Queen (*Stichophthalma camadeva*) - NE Hills, Sikkim
Orange albatross (*Appias nero*)
Owl moth (*Brahmaea wallichii*)
Papilio castor ssp. *polias* - E. Himalaya
Papilio krishna krishna - E. Himalaya
Paris Peacock (*Papilio paris*)
Psyche butterfly (*Leptosia nina*)
Red lacewing (*Cethosia biblis*) NE India
Striped blue crow (*Euploea mulciber*)
Tawny rajah (*Charaes bernardus*)
Teinopalpus imperialis - E. Himalaya
Tufted jungle Queen (*Thauria* spp.)
Vanessa vau-album - NW Himalaya (found along the Timber line).
Yellow coster (*Acraea vesta*)
Ypthima methora (Great Five Ring)- Assam and Sikkim

Ethnobiological data on biodiversity

Though the present century is the day of synthetics and other modern products but the importance of lore of primitive and indigenous communities has not diminished. More and more drugs are being discovered from wild plants. The ethnic groups are rather isolated from one another and practice agriculture, selection of species for use and maintain the systems for their welfare. The intimate relationship of ethnic groups with the forests have helped in the conservation of many species, their knowledge about the flora and fauna, its diversity and use have been widely acclaimed. The primitive knowledge of about 3000 plant species that have been used for various purposes including population control has been gathered, screened and computerized. The WHO (World Health Organization) has started documentation and testing of popular tribal remedies (Biswas, 1980, 1990).

Ethnobiological studies reveal that out of nearly 10,000 wild plant species, 7500 bear medicinal property, 3900 edible, 525 fibre and cordage yielding, 400 used as fodder and 300 contain piscidal and pesticidal elements. Among the fauna 76 species are the source of ethnic medicine, the former representing 60 vertebrates and 16 invertebrates (AICPE, 1999). Since these species of ethnobiological significance are in specific milieu with the communities, therefore, they need the attention for conservation and sustainable utilization.

Conservation Strategies

The different kinds of protective measures related to conservation can be grouped into two main categories: *in situ* and *ex situ* conservation.

For long-term conservation, the conservation of genetic resources through their maintenance within natural ecosystem is the ideal system, e.g. comprehensive system of protected areas. The different categories of these include National Parks, Sanctuaries, Nature Reserves, Biosphere Reserves, Gene Sanctuaries and several others. There are 89 National Parks and 489 Wildlife Sanctuaries covering 154, 572.80 sq km or 4.70% of the country's geographical area. Some of the states and Union Territories of India such as Punjab, Tripura, Chandigarh, Dadra & Nagar Haveli, Lakshadweep, Pondicherry, Daman & Diu have no National Parks. Best covered area from total protected area point of view are Sikkim (28.88%), Goa (20.39%), Himachal Pradesh (12.93%), Uttaranchal (12.10%), Arunachal Pradesh (11.82%), Gujarat (8.62%), Kerala (6.90%), Jammu & Kashmir (6.69%), Tripura (5.76%) and Orissa (5.1%), (Rodgers *et al.*, 2002). Besides, there are sacred groves being maintained on religious grounds. Biosphere Reserve network programme was launched by UNESCO (United Nation Educational, Scientific and Culture Organization) in 1974 under its Man and Biosphere Programme with the main objectives to conserve representative samples of ecosystem, to provide long-term *in situ* conservation of genetic diversity and to promote basic and applied research and monitoring besides facilities for education, awareness and training. There are 356 Biosphere Reserves in 90 countries by the year 1998. Under Indian Wildlife (Protection) Act of 1972 the country has 23 Tiger reserves.

People's participation in biodiversity conservation is necessary as also the protection of the protected areas. Role of Armed Forces (Example: Ecotask Force activities) in this regard has been very significant in recent times.

Following 12 sites have been established as Biosphere Reserves by the National MAB committee under the aegis of the Department of Science & Technology and Environment, Government of India: Nilgiri (Tamil Nadu, Karnataka & Kerala), Gulf of Mannar (Tamil Nadu), Nokrek (Tura Ridge), Meghalaya, Great Nicobar (Andaman and Nicobar), Nanda Devi (Uttaranchal), Sunderbans (West Bengal), Manas (Assam). Simplipal (Orissa), Dibru Saikhowa (Assam), Pachmarhi (Madhya Pradesh), Kangchedozonga (Sikkim) and Dihang-Dibang (Arunachal Pradesh). Few more Biosphere Reserves (BR) have been proposed such as Cold Desert BR, Little Rann of Kuchch, Amarkantak BR etc. (see Maikhuri *et al.*

S. No.	Biosphere Reserves	States	Year (Est.)	Biogeographic Zone
1.	Nilgiri	TN,KNTK,KRL	1986	Deccan Peninsula
2.	Nanda Devi	Uttaranchal	1988	W. Himalaya
3.	Nokrek	Meghalaya	1988	NE India
4.	Great Nicobar	A & N Islands	1989	Islands
5.	Gulf of Mannar	TN	1989	Coastal
6.	Manas	Assam	1989	NE India
7.	Sunderbans	West Bengal	1989	Gangetic Plain
8.	Similipal	Orissa	1994	Deccan Peninsula
9.	Dibru Saikhowa	Assam	1997	NE India
10.	Dihang-Dibang	Arunachal Pradesh	1998	NE India
11.	Pachmarhi	Madhya Pradesh	1999	Central Highland
12.	Kangchedozonga	Sikkim	2002	East Himalaya

Abbreviation: KNTK = Karnataka, KRL = Kerala, TN = Tamil Nadu

1998; Biswas 2004). These cover different biogeographic regions of the Indian subcontinent. The different biogeographic areas are: Himalayan Highlands, Thar Desert, Malabar Rain Forest, Deccan Thorn Forest, Coromandel Mahanadian, Laccadive (Lakshadeep) Islands, Burman Monsoon Forests, Bengalian Rain Forests, Maldive and Chagos Islands, Andaman and Nicobar Islands etc.

Rodgers *et al.* (2002) while emphasizing the conservation significance of mammal species reported that out of 18 species listed in 1988 presently have no or insignificant population in a National Park. The species are Indian Wild Ass (Kuchch), Phyris Leaf Monkey (Assam Hills - Tripura), Malabar Civet (Western Ghats, K erala), Malay Sun Bear (Arunachal Pradesh, Mizoram), Markhor (NW Himalaya, Jammu & Kashmir), Argali / Nayan (Ladakh), Pallas' Cat (Ladakh), Rusty spotted Cat (Maharashtra & Gujarat), Tibetan Antelope (Ladakh, Sikkim), Tibetan Gazelle (Ladakh), Pig tailed Macaque (Assam Hills) and Stump tailed Macaque (Assam Hills).

The floristic and faunistic explorations in un/underexplored tracts and ranges are required to be intensified and the components under threat of extinction, also the spectacular flora and fauna of great biological interest, horticultural importance and primitive flora of antiquity should be protected *in-situ* and *ex-situ*. S etting-up of Environmental Centres (Paryavaran Kendras in each Biosphere Reserve for undertaking research and training work on different aspects are suggested. These centers can also play important role in educating students and people about the environment, forest and wildlife. Armed Forces may adopt certain biodiversity - specific areas and also the region-wise species for preservation and conservation. While suggesting the environmental priorities in India and sustainable development based on works done by many organizations Dr. Khoshoo, an eminent naturalist and biologist in his Presidential Address of Indian Science Congress Association held at New Delhi in 1986 suggested state- specific species of mammal, bird, wild flower and trees for particular state of the country for conservation purpose as following:

Andhra Pradesh	:	Four-horned antelope, Grey Pelican; Jasmine (*J. pubescens*), Red Sanders.
Arunachal Pradesh	:	Takin, Peacock, Pheasant, Fox-tail, Orchid, Hollong (*Dipterocarpus macrocarpus*).
Assam	:	One-horned Rhino, White-winged wood duck, Champa (*Michelia champaca*), Bamboo (*Arundinaria griffithiana*).
Bihar (including Jharkhand)	:	Sloth Bear, Comb duck (Nakta), Kachnar (*Bauhinia purpurea*), Mahua (*Madhuca indica*).
Delhi	:	Hanuman Langur, (*Presbytis entellus*), Grey Partridge (*Francolinus pondicerianus*), *Lagerstroemia indica,* Arjun (*Terminalia arjuna*).

Goa	:	Mouse Deer (*Muntiacus muntiak*), Brownheaded Sea Gull (*Larux brunnicephalus*), Wild Morning Glory (*Ipomea pes-caprae*), *Hibiscus tiliaceus* (Tulip Tree).
Gujarat	:	Asiatic Lion, Flamingo, Amaltas (*Cassia fistula*), Neem.
Haryana	:	Nilgai, Black Partridge, Vasak (*Adhatoda vasica*), Babul (*Acacia nilotica*).
Himachal Pradesh	:	Musk Deer, Monal Pheasant, Atis or Bish (*Aconitum heterophyllum*), Deodar.
Jammu & Kashmir	:	Kashmir Stag (Hangul), Western Tragopan, (*Tragopan melanocephalus*), Blue Poppy (*Meconopsis aculeata*), Horse Chestnut (*Aesculus indica*).
Karnataka	:	Slender Loris (*Loris terdigradus*), Great Pied Hornbill (*Buceros bicornis*), Nandivardhan (*Ervatamia coronaria*), Sandal Wood (*Santalum album*).
Kerala	:	Lion-tailed Macaque (*Macaca silenus*), Large Racket - Tailed Drongo (*Dicrurus paradiseus*), Scarlet *Ixora* or Techi (*Ixora coccinea*), Rosewood (*Dalbergia latifolia*).
Madhya Pradesh	:	Barasingha (*Cervus duvauceli*), Paradise Flycatcher, *Sterculia (Sterculia colorata),* Banyan tree (*Ficus benghalensis*).
Maharashtra	:	Gaur (*Bos gaurus*), Grey Jungle Fowl (*Gallus sonneratii*) Jarul (*Lagerstroemia speciosa*), Teak (*Tectona grandis*).
Manipur	:	Thamin or Sangai (*Cervus eldi* in Keibul Lamjao), Hume's bar Backed Pheasant (*Syrmaticus humiae*), Manipur Lily (*Lilium macklinae*), Toon tree (*Toona ciliata*).
Meghalaya	:	Binturong (*Arctotis binturong*), Hill Myna (*Gracula religiosa*), Pitcher Plant (*Nepenthes khasiana*), Wild Orange (*Citrus hystrix*).
Nagaland	:	Clouded Leopard (*Neofelis nebulosa*), Imperial Pigeon (*Muscadirora aenea*), Blue vanda (*Vanda coerulea*), Areca Nut (*Areca catechu*).
Orissa	:	Sambar, (*Cervus unicolour*), Blue Jay (*Coracias benghalensis*), Ashoka (*Saraca asoca*), Tendu (*Diospyros melanoxylon*).

Punjab	:	Blackbuck (*Antelope cervicapra*), Hoopoe (*Upupa epops*), Coral tree (*Erythrina variegata*), Shisham (*Dalbergia sissoo*).
Rajasthan	:	Gazelle, (*Gazella dorcas*), Great Indian Bustard (*Ardeotis nigriceps*), Karil (*Capparis deciduas*), Khejri (*Prosopis cineraria*).
Sikkim	:	Red Panda (*Ailurus fulgens*), Blood Pheasant (*Ithaginis cruentus*), Noble orchid (*Dendrobium nobile*), *Rhododendron thomsonii*.
Tamil Nadu	:	Nilgiri Tahr, (*Hemitragus hylocrius*), Spoon-bill (*Platalea leucorodia*), Pavalakkurinja (*Crossandra infundibuliformis*), Tamarind (*Tamarindus indicus*).
Tripura	:	Phayer's Leaf Monkey (*Presbytis phayrei*), Fairy Blue Bird (*Irena puclla*), Nagkesar (*Mesua ferrea*), Agar (*Aquilaria malaccensis* syn. *A. agallocha*).
Uttar Pradesh	:	Leopard (*Panthera pardus*), Sarus crane (*Grus antigone*), Sal (*Shorea robusta*).
Uttaranchal	:	State Government has adopted Musk deer as its State animal. *Rhododendron arboreum* (Burans) as its tree, monal as bird and *Saussurea obvallata* (Brahmkamal) its flower. (*Hindustan Times* 1st Oct. 2001).
West Bengal	:	Indian Elephant (*Elephas maximus*), Barn Owl (*Tyto alba*), Sheuli (*Nyctanthes arbor-tristis*), *Bischofia javanica*.
Andaman & Nicobar	:	Crabeating (*Macaca fascicularis*), Megapode (*Megapodius freycinet*), Keya (*Pandanus andamanicus*), Padauk (*Pterocarpus dalbergioides*).
Mizoram	:	Golden Cat (*Felis temmincki*), Blyth's Tragopan (*Tragopan blythii*), Chaalta (*Dellenia indica*), Mizo Champa (*Michelia montana*).

Ex-situ conservation is conservation outside the natural habitat of species and is possible by perpetuating sample populations in genetic resource centers, botanical gardens, tissue culture laboratories, germplasm banks for clonal materials, seeds, pollen etc.

The country has 64 main and 194 small or medium zoos. These zoological gardens are represented by numerous rare and threatened fauna and its population. It has been suggested time to time that the over populated species of endangered groups of animals should be released to their original habitat and region of distribution.

Regionwise Brief List of Spectacular, Endemic Rare and Endangered Species of Plants for Conservation

Jammu and Kashmir	:	*Lavatera kashmeriana, Cyperipedium cordigerum, Saussurea lappa.*
Himachal Pradesh	:	*Dioscorea deltoidea, Balanophora involucrata.*
Uttar Pradesh & Uttaranchal	:	*Olax nana, Eremostachys superba, Saussurea* spp., *Mahonia jaunsarensis, Sophora mollis, Hovenia dulcis, Acer species, Taxus baccata.*
Rajasthan, Gujarat & Gangetic Plains.	:	*Rosa lyellii* (Mt. Abu), *Tecomella undulata, Hyphne dichotoma.*
N.E. India	:	Orchids (*Vanda, Paphiopedilum* i.e. Lady's slipper, *Cymbidium, Dendrobium* etc.). *Taxus baccata, Amentotaxus assamica, Magnolia pterocarpa, Nepenthes khasiana* (Pitcher Plant), *Sapria himalayana* (Flower 15-35 cm in diam.) *Populus gamblei, Panax pseudoginseng, Rhododendron species, Dischidia rafflesiana, Helwingia himalaica* (Flowers on the midrib of leaf), *Coptis teeta*, Blue poppy (*Meconopsis* sps), *Primulas, Calamus inermis.*

Takhtajan (1969) while delving into the origin of species named following species of North-eastern India and Myanmar as primitive flowering plants, namely *Magnolia (M. griffithii, M. pealiana, M. gustavii, M. pterocarpa*), *Manglietia, Euptelea, Tetracentron, Pycnarrhena, Haematocarpus, Aspidocarya decaisnea, Holboellia, Stauntiana, Parvatica, Exbucklandia, Distylium, Altingia, Houttuynia, Myrica esculenta, Alnus* and *Betula* having phytogeographical distinctiveness.

Peninsular India	:	Goa; *Entada pursaetha, Angiopteris evecta.*
Andhra Pradesh	:	*Cycas beddomei, Pterocarpus santalinus.*
Tamil Nadu	:	*Rhododendron nilagiricum, Lilium neilgherense, Podocarpus wallichiana* (South Indian Conifer).

Karnataka & Maharashtra	:	*Ceropegia* species, *Viscum mysorense*. Reed Bamboo (*Ochlandra* spp.), *Abutilon ranadae* etc.
Kerala	:	*Bentickia coddapana, Hopea utilis, Paphiopedilum druryi, Toxocarpus palghatensis* etc.
Andaman &	:	*Ailanthes kurzii, Canarium mannii, Dipterocarpus*
Nicobars	:	*kerrii, Hopea andamanica, Lagerstroemia hypoleuca, Calamus dilaceratus, Korthalsia rogersii* etc.

Needs and Priorities for Sustainable Utilization and Conservation of Endangered Biodiversity

1. Surveys - Floristic, Faunistic and Ethnobiological in threatened areas rich in biodiversity and genetic resource. Study on traditional methods of selection, cultivation, sustainable utilization, domestication, rearing and preservation.

2. Application of recent trends in habitat evaluation, classifying, evaluating and integrating the components. (Remotely sensed/Aerospace data, GPS & GIS database, Biochemical analysis to widen the genetic base, Palynology, Cytogenetics, Chemotaxonomy and Taximetrics).

3. Developing appropriate multiplication, harvesting methods, grading and utilization technology for the rare species being exploited for economic purposes.

4. People participation through different dimensions of joint forest management (JFM) and other social forestry programmes. Involving people in management by deriving economic and commercial benefits for them.

5. Emphasis on *in situ* conservation and rehabilitation/release of artificially developed population /reproductive materials in congenial areas. Over crowded zoos with rare and threatened fauna should be made spacious by releasing the population into original areas of their occurrence. There is need for zoo-cum-botanic gardens representing the threatened groups of plants and animals.

6. Monitoring and reassessment of different measures undertaken for the conservation of threatened biodiversity.

7. Media orientation and coverage on significance of biodiversity and inclusion of subject at schools to university levels.

Literature Cited

AICPE (All India Co-ordinated Project on Ethnobiology) (1999): Ethnobiology in India - a status report - AICPE. Ministry of Environment & Forests, New Delhi.

Alfred, JRB; Kumar, A; Tak, PC and Sati, JP (2001): Waterbirds of Northern India Zoological Survey of India, Govt. of Inda, Kolkata - 700 020.

Biswas, Sas. (1980): Endangered plant heritage. In: Sharat Chandra, Joseph & Goswami (Eds.) "Development without Destruction". *Proceed.* of Symposium 3-4th Oct. 1980: 10-13. Forest Department, Govt. of Meghalaya.

Biswas, Sas. (1990): Ethnobotanical studies on North-eastern region of India with particular emphasis on conservation of biological diversity. *Proceed. 2nd International Congress of Ethnobiology. "The Challenges of Ethnobiology in the 21st Century.* 22-26 Oct. (Abs.) Kunming, P.R. China: 128.

Biswas, Sas. and Dayal, R. (1994): Endangered biodiversity of Indian flora and strategies for its conservation In: Sahni, K.C. (Ed.) "*Advances in Plant Science Research*" Vol. II: 73-91. IBD, Dehra Dun.

Biswas, Sas. (1995): Phytogeographical studies on the forest flora of Tehri Garhwal Himalaya (Uttar Pradesh) In: Gupta, B.K. (Ed.) *Higher Flants of India Subcontinent* 1: 223-234. BSMPS. Dehra Dun.

Biswas, Sas. (1995a): Biodiversity indischer flora (Kurzfassung). Paper presented in Institutional Seminar of Institut fur Forstgenetiks, Grosshansdorf, Federal Republic of Germany: 1-5.

Biswas, Sas.; Chandra, S. and Jain, S.S. (1996): Rare and endangered bamboos rattans and medicinal plants of Indian Himalaya and strategies for their conservation. *In*: Karki, M. *et al.* (Eds.). *Proceed. International Workshop on Role of Bamboos, Rattan and Medicinal Plants in Mountain Development* Pokhara, Nepal May, 15th-18th, 1996: 113-121.

Biswas, Sas. and SS Jain (2002): Is Copeland's bamboo (*Bambusa copelandi* Gamble ex Brandis) 'presumed extinct'? - A taxonomic search for its whereabouts *Indian J. Forestry* 25 (4):472-475.

Biswas, Sas. (2004): *In situ* conservation - a multidisciplinary participatory task. *In* Dhillon, BS *et al.* (Eds.) Proceed. of symposium on "*Plant Genetic Resource Management*": 205-219. Narosa Pub. House, New Delhi.

Grimmett, R.; Inskipp, C. and Inskipp, T. (1999). Pocket guide to the birds of the Indian subcontinent, Oxford University Press, New Delhi.

Groombridge, B. (1992): Global biodiversity-status of earth's living resources - a report compiled by WC Monitor Centre: 78, 239. Chapman & Hall, London.

Hanski, I., Clobert, J. and Reid, W. (1995): Generation, maintenance and loss of biodiversity. *In*: Heywood, VH and Watson, RT (Eds.) *Global biodiversity assessment*. Cambridge University Press. UNEP.

Harper, J.L. and Hawksworth, D.L. (1996): Preface. In: Hawksworth, D.L. (Ed.). *Biodiversity - measurement and estimation:* 13. Chapman & Hall, London/ Madras.

Joshi, K.D. (1999): Piscine diversity in Kumaon Rivers (Central Himalaya). *J. Inland Fish. Soc. India* 31(2): 20-24.

Khoshoo, T.N. (1986): Environmental priorities in India and sustainable development: 75-101. Indian Science Congress Association, Calcutta.

Kushwaha, SPS (1997): Landmass dynamics and rhinohabitat suitability analysis in Kaziranga National Park, Assam - Project Report, IIRS, Dehra Dun.

Maikhuri, RK; Rao, KS and Rai, RK (Eds.) (1998): Biosphere reserves and management in India. Himavikas Occassional Pub. No. 12. GB Pant Institute of Himalayan Environment & Development, Kosi-Katarmal, Almora.

Mani, G.S. (ed.) (1974): Ecology and biogeography in India. pp. 773. W. Junk. Hague.

Rodgers, W. Alan; Panwar, HS and Mathur, VB (2002): Wildlife Protected area network in India: a review (Executive Summary). Wildlife Institute of India.

Salim Ali (1964): The book of Indian birds. Bombay Natural History Society, Bombay.

Swaminathan, MS (1993): Agriculture cannot wait. *Everyman's Science*. 28 (I): 17-24.

Takhtajan, A. (1969): Flowering plants: Origin and dispersal (Translated by C. Jeffrey). Oliver & Boyd. Edinburgh.

Tewari, D.N. (1994): Biodiversity and forest genetic resources: 251-311. IBD, Dehra Dun.

Wildlife (Protection) Act (1972): The Wildlife (Protection) Act, 1972 as amended in 2001. Universal Law Publicity Co. Pvt. Ltd.

Vavilov, N.I. (1951): The origin, variation, immunity and breeding of cultivated plants. *Chronica Botanica* 13: 1-364.

Chapter II

Under Exploited Forest Biodiversity for the Environment of Present Millennium

The sky-high development of mankind during the bygone century is at the cost of environmental degradation and loss of unexplored biodiversity and genetic resource. The better living conditions rested upon urbanization, industrialization, mechanization-cum-modernization etc. The environmental priorities for the millennium draw attention on the issues such as land degradation, climate change, acid rains, nuclear winters, aridity, local pollution, ozone layer rejuvenation, biodiversity conservation and the sustainable utilization of exhaustible natural resource. It is a striking fact that the renewal of resource depends upon sustainable utilization, effective enforcement of conservation measures and checking up further erosion of potential biodiversity.

In terms of species diversity of natural resource and the quantum of well-known species meeting the need to fulfill the lacunae in the knowledge of elements for multifarious utilization by the mankind and subsequently ameliorating the environment is limited. Following aspects are emphasized for better understanding (Biswas, 2000):

- lesser-known species in well-known environment
- well-known species in lesser-known environment
- well-known species in well known/worked out environment

The criteria for the selection of species, environmental priorities, components of environment (habitat, ecosystem etc.) and factors responsible for the introduction of species cover any of the aforementioned aspects.

The present millennium opens the new frontiers for the integration of advances made in the field of science and technology (S&T) in order to achieve the common goal for a better environment. The latter encompasses the following main areas with respect to the lesser known or secondary species of immense potential.

- species for all kinds of pollution (air, water, noise, indoor).
- species for the environment friendly and active principles (medicine, biopesticides, dyes etc.)
- species for food and nutrition.
- species for commercial utilization (industry, construction or settlement etc.) and alternate source of conventional wood resource.
- species for protecting disaster-prone areas (seismic, cyclonic impacts, tsunami etc.)
- species for conserving and rehabilitating the fragile areas of different ecosystem (mountain/desert/aquatic/mangrove).
- species for energy plantation (fuel, petrochemicals or hydrocarbon).
- species for urban bioaesthetics and landscaping.
- species to replenish the lost mosaic of biodiversity (*in situ*, reintroduction, relocation).
- integrating indigenous knowledge (IK) of lesser-known elements with recent advances of S & T.
- species for specific agrosystem.
- development of database on potential species of environmental and biodiversity conservation.
- ecological assessment and economic evaluation of species.

The potential species of cultigens are required to be evaluated with a view to deciphering the infra specific traits and intermediate forms of environmental significance. Some of such species introduced in zones that are presently environmentally - stressed have acquired certain characteristics indicative of a particular influence of environment it has undergone.

The past millennium is referred as the period of accumulation of knowledge on the inventory of nature's resources. The data gathered was subjected to the conversion from a *status of sterility to fertility*. The population of India is expected to touch a figure of 1013 million by first few years of this century. It means that demand for various values - added products from nature would subsequently increase. Besides, the number of species growing under stressed conditions would increase due to increase in influencing factors.

Choice of Potential Species to Combat Pollution

It is well known that vegetation ameliorates pollution most effectively through its ability to reduce airborne particulate, break wind speed and absorb noise. Some of the striking examples are that of Douglas fir with 30 cm diameter of trunk which removes 60 mg of cadmium, 140 mg of chromium, 820 mg of nickel, and 5200 mg of lead in Connecticut, USA. The plants reduce normal levels of benzene 50-90 percent within 24 hours and trichloroethylene by 20 percent; golden pothos, a lesser-known plant reduces 100 times the normal level of aforementioned gaseous pollutants (Lanphear 1971, Bernatzky 1978, Schmid 1975, Wolverton *et al.* 1989). The urban world of several of developing nations with automobiles running over the carrying capacity of roads and with no check on pollution may experience in early of this century hazardous effects of air pollution. Several of the developed metros of the world where there are active measures over polluting agents have 3000 particulate per litre of air along the main streets with trees compared to the sites without vegetation containing 10000-12000 particulate e.g. Frankfurt. Bhattacharya (1994) enlisted pollutant (sulphor dioxide, fluorides, hydrogen sulfide, mercuric vapours ethylene etc.) sensitive and resistant species citing largely the exotic elements of European and American origin.

Studies on the impact of pollution on Indian forest trees typical of urban areas with a view to selecting the resistant species have not been by and large carried out. Infact, for such species as in teak (*Tectona grandis*) an annual loss of wood production through decrease in the quantum of radial growth is reported by Ghouse *et al.*(1984) and it is found to be 26 percent on an average of 6 years of observation over 30 years old teak trees growing in the vicinity of a Thermal Power Station. A systematic analysis of the impact of pollutants on different morphological phenotypes within the species may reveal useful data on the types resistant to the pollution. The resistant forms may subsequently be introduced in the stressed sites.

Potential species to help combating pollution are suggested under following major plant groups:

- Gymnosperms (*Abies, Picea, Pinus, Cedar, Cypress, Thuja, Larix, Ginkgo, Junipers*).
- Ash (*Fraxinus*)
- Poplars and willows (*Populus, Salix*)
- Maples (*Acer*)
- Myrobalans (*Terminalia*) and laurels.
- Mesquites and other legumes (*Prosopis, Gleditsea, Sesbania*)
- Oaks (*Quercus, Lithocarpus* etc.)
- Birch (*Betula*) and beech (*Fagus*).
- Meliaceae (broad - leaved species with compact crowns)
- Elms (*Ulmus* spp.)
- Horse Chestnut (*Aesculus, Hippocastanum*)
- Myrtaceous species (*Eugenia, Syzygium,* Eucalypts)
- European Plane (*Platanus* spp.)
- *Rosaceae* (*Sorbus, Prunus, Pyrus*)

Choice of Potential Species of Medicinal Value for the Millennium

Among the medicinal and aromatic plants of environmentally sensitive himalayan ecosystem, the lesser-known species as indicated after the following species in parenthesis is quite appreciable e.g. *Aconitum* (25), *Artemisia* (20), *Berberis* (20), *Capparis* (4), *Colchicum* (91), *Coleus* (1), *Costus* (6), *Dioscorea* (5), *Gentiana* (50), *Inula* (12), *Jurinea* (2), *Panax* (2), *Saussurea* (26), *Valeriana* (8) etc. (Biswas *et al.* 2000). As per a study (Myers 1983) made on a country-by-country basis through a small-scale screening of a unevenly distributed proportion of different plant species, it is revealed that plants of Indian flora bear 4.8 percent anticancer activity. Among the plant groups (families) little and less-known species of potential value are represented by taxa such as *Apocynaceae, Araceae, Dioscoreaceae, Lauraceae, Leguminosae (sensu lato) Rutaceae, Zingiberaceae* etc. For some dreadful diseases such as AIDS, HIV etc. which appeared towards the later part of the last century have answer for their eradication hidden in species of forest origin. Species such as *Phyllanthus nirurii,* a complex of several species, *Emblica officinalis, Castanospermum australe* etc. are some highly significant examples. Ethnobiological information on nearly 80 species of forest fauna for medicinal value exists and needs further investigation for wider and restricted use.

Conservation Efforts and High Population Growth

India's population, which was 846.3 million in 1991 became one billion on 11th May 2000 and expected to escalate to a mammoth 1236.5 million in 2016. National Population Policy (NPP)-2000 apprehended that the country's population is highly alarming if its current annual rate of 15.5 million continued unabated (*Hindustan Times,* dated 2nd April, 2000).

Ever exploding population pressure of the country has drawn the attention of concerned government machinery and authority to introduce among the people safe and secure plant-based medicines of contraceptive principles. The cause of the population growth rate is attributed to the *unrealized need for contraception to young couples* as per National Population Policy (NPP)-2000. Among the plant-based prescriptions only limited species are well known e.g. *Dioscorea deltoidea, Costus speciosus, Sapindus mukorosii* (CONSAP) and few more. But the active principle occurs in several species, the knowledge about which is largely embedded in traditional systems of different indigenous communities (Biswas *et al.* 2000).

Potential Species of Biopesticidal Value

Synthetic insecticides and pesticides manufactured from chemical base have adverse impact on the environment. Besides, the insects develop resistance on sustained use of insecticides. As per a report, India loses grain worth over Rs.5000 million each year due to insect pest (Tewari, 1992). Keeping the congenial environment in mind, biopesticides upsetting the insects' hormonal balance and making the pests' antifeedant are highly significant. Approximately 2500 plants belonging to nearly 65 families possess biopesticidal properties. Among the lesser-known plant groups with under-explored forest species *Apocynaceae, Asclepiadaceae, Asteraceae, Borginaceae, Burseraceae, Clusiaceae, Ebenaceae, Fabaceae, Flacourtiaceae, Hippocastanaceae, Lamiaceae, Lauraceae, Magnoliaceae, Meliaceae, Myrsticaceae, Myrtaceae, Rosaceae,*

Sapindaceae, Simaroubaceae, Theaceae, Verbenaceae etc. deserve specific attention.

Potential Forest Species for Food and Nutrition to keep the Pressure off from Conventional Resource

The forest species with value - added food and nutrition are nearly 2000 which occur in different bioclimatic and agroclimatic zones (Bennet, Sas. Biswas *et al.* 1991). Several of the species do not only occur in ecologically fragile region but they are lean season food for the dependents also. The species and the races with appreciable food and nutritive values being cultivated and sustainably utilized are the result of several decades of selection by man subconsciously/consciously and through permutation and combination.

The potential of forest species with food and nutritional value is enormous but remains by and large under-utilized. It is estimated that 80 percent of the forest dwellers in Bihar, Madhya Pradesh, Orissa and Himachal Pradesh depend on forests for 25-50 percent of their annual food requirements (CSE, 1985). Due to stressed environment and scarcity of variety of food of forest origin some of indigenous communities, for example in Kalahandi environ of Orissa succumbed to hunger, and malnutrition and infected food. The nutritional value of only very well utilized crop plants is known whereas that of forest species has long been least concerned. Since the present millennium would see the dependence of man to greater extent on un/under - utilized forest species more nutritional than food value. Although over 100 species of forest fungi are edible but only about 25 species have become palatable (AN Shukla, *pers. comm.*). There is urgent need for characterizing more and more potential species of forest origin. The land races among certain forest species known by their nutritional value (Protein) - *Balanites aegyptiaca* (50%) in se ed kernel, *Sterculia urens* (35%), *Terminalia catappa* (25.42%), *Pithecellobium dulce* (17.6%), *Holoptelea integrifolia* (13.7%), *Aisandra butyracea* (7.10%), *Feronia limonia* (7.3%), etc. are needed to be explored for higher level of quality and quantity of protein. Similarly for the carbohydrate the various forest species known for the quantity as mentioned here e.g. *Dioscorea* spp. (71-85%), *Holoptelea integrifolia* (69.91%), *Ficus racemosa* (49%), *Cochlospermum religiosum* (35.78%), *Phoenix sylvestris* (33.8%), *Aegle marmelos* (31.8%),

Manilkara hexandra (27.74%), *Flacourtia ramontchii* (24.20%), *Salvadora oleoides* (23.48%), *Aisandra butyracea* (21.79%), *Pithecellobium dulce* (19.9%), *Catunaregum spinosa* (17.7%) etc. are needed to be studied for their intermediate/infra specific forms for better yield and potential for genetic improvement.

Among the lesser-known species based on several decades of floristic, ethnobiological and other surveys following posseses enormous potential for the present millennium (*vide* Biswas 2000; Biswas *et al.* 2000; Bennet, Biswas *et al.* 1991).

Alpinia galanga, A. zerumbet, Amaranthus caudatus, A. graceizans, A. spinosus, A. tricolor, A. viridis, Amorphophallus bulbifer, A. commutatus, A. dubius, A. paeoniifolius, A. sylvaticus etc., *Arenga saccharifera, Antidesma* species, *Aponogeton* spp., lesser-known species of *Artocarpus* such as *A. hirsutus, A. nitidus* etc.

The species of bamboo being used as food by several indigenous communities deserve conservation. The gregarious flowering of this monocarpic species sometime results into the decline of its resource.

Caryota mitis, C. urens, Castanopsis species, *Ceropegia* (25-30 spp.). The tuberous root has immense potential; *Coix lachryma-jobi*; *Cordia* spp., *Corypha umbraculifera, C. utan*; Cyperaceae (Cyperus spp.), *Dioscorea alata, D. belophylla, D. bulbifera, D. esculenta, D. hamiltonii, D. hispida, D. melanophyma, D. oppositaefolia, D. puber, D. wallichii* etc. *Diospyros* spp., *Diplazium asperum, D. esculentum; Docynia hookeriana, D. indica, Elaeocarpus floribundus, E. lanciaefolius, E. serratus, Euryale ferox* - an aquatic food crop, *Emblica officinalis* and its variability, *Ficus* spp. (25 spp.). The polymorphic, phenologically variable and environmentally specific species need characterization. *Hodgsonia macrocarpa*: the kernel i.e. 33 percent of whole seed contains protein upto 21.5% and fatty oil 66.5%, the latter being non drying has cooking potential. *Houttuynia cordata, Impatiens* (wild spp.), *Lasia spinosa, Maclura cochichinensis, Mangifera andamanica, M. sylvatica, M. camptosperma, M. cacharensis* are all lesser-known species.

Promising Under-utilized Plants of Food Value

Maughania vestita, a lean season tuber crop of indigenous community of Meghalaya; *Moringa concanensis, M. oleifera, Musa* (Wild species and land races); *Nelumbo nucifera, Nymphea alba, N. nouchali, N. stellata; Nymphoides indica, Nypa fruticans*, a palm of inundated coastal areas of the country; *Pandanus* spp. (*P. andamanesium, P. fascicularis, P. leram*), *Parkia timoriana, Phoenix acaulis, P. humilis, P. pusilla, P. sylvestris, Pithecellobium dulce, Polygonatum* (8 species), *Polygonum* (9 spp.), *Prosopis cineraria* and its variability occurring in arid and semi arid areas of the country; *Prunus armeniaca* - The fruit pulp of the species could be found as a nutritious food item of lean season of Himalayan region; *Pueraria tuberosa, Pyrus aucuparia, P. pashia, Quercus* species (Oaks *sensu lato*); *Rheum australe, R. speciforme, R. webbianum, Rhizophora mucronata* and related species, *Ribes species, Rorippa nasturtium - aquaticum, R. aquaticum, Rosa gigantea, R. sericea, R. webbiana, Rhizophora mucronata* and related species *Rhodomyrtus tomentosa* var. *parviflora; Rubus* species, *Rumex acetosa, R. maritimus, Salvadora oleoides* and *S. persica, Sauropus androgynous* referred popularly as '*multivitamin green*' has composition with Mineral matter 3.4%; Calcium 0.57; Phosphorus 0.20; Iron 28.0 mg; Carotene (Vitamin A); 9510 I.U.: Thaimine 0.48; Riboflavin 0.32; Nicotine acid 2.6; and Vitamin C 247 mg/ 100 gm; Moisture73.6%; Protein 6.8%; Carbohydrates 11.6% and Fibre 1.4% per 100 gm. (see Bennet, Biswas *et al.* 1991).

Various species of sal, namely (*Shorea robusta, S. roxburghii, S. thunbaeggia, S. retusa* syn. *S. assamica*) contain seed meal of food value. Other species are *Spondias pinnata* and its close allies, *Syzygium* species, *Tacca leontopetaloides, Vaccinum species, Wallichia disticha, Ximenia americana* and *Zingiber zerumbet.*

Lichens, Algae and Fungi - Food for Present Millennium

While the occurrence of lichens are indicators of pollution free environment, several of them are edible, for example, *Alectoria jubata, Evernia prunastri, Lecanora esculenta, Peltigera canina, Ramalina sinensis* and *Usnea longassima*. Among the ferns, *Angiopteris evecta* (rhizome), *Asplenium* species, *Ceraatopteris setiquosa, Diplazium esculentum, Equisetum debile, Marsilea minuta, Pteridium aquilinum* (rhizome), are some of edible examples (Arora and Pandey 1996). The potential edible algae for the present millennium are *Chlorella pyrenoidosa, Chondrus crispus, Codium* species, *Coelastrum probascideum* (high protein), *Gracilaria* spp., *Hydrodictyon reticulatum* (fish like small), *Hyphnal musciforrmis* (palatable jelly), *Laurlnicia obtusa, Nostoc commune* (as delicacy), *Porphyra vietnamensis, Scenedesmus acutus* and *Spirulina plaatensis* (high protein), *Turbinaria* and *Ulva* spp. Several of the diversity of algae and lichens occur in a typical forest ecosystem. The species of fungi such as *Agaricus campestris, Auricularia sambucina, Cantharellus cibarius, Clavaria* spp., *Clitocybe* spp., *Collybia* spp., *Fomes pachyphlaes, Hydnum* spp., *Lentinellus cochleatus, Lentinula edodes, Morchella esculenta, Podebrella microcarpa* etc. found in natural and modified ecosystems are edible.

Potential Species of Fuel Energy

With the depleting resource of fossil fuel and petroleum reserves the new millennium has to find an alternative source in plant world. World over there is exploration for alternate conventional and non conventional source of energy for the clues about the existence of crude petroleum as bio-fuels appeared as a consequence of man's pursuit to find a substitute in times of emergent need. The '*petroleum nut tree*' of the Philippines (*Pittosporum resiniferum*) was exploited by the Japanese to obtain tank fuel during World War II, by virtue of its highly volatile oil. According to Myers (1983) merely six well-grown trees of this species can produce 300 liters (about 70 gallons) of oil per year for variety of uses. Phenologically the tree produces fruits twice in a year and one kg of nuts when pressed, yield about 70 gms of oil. The latter burns with a smokey flame at 300 degree celsius. It is reported that freshly picked nuts can be lighted with a match. Another striking example of '*phytopetroleum*' yielding species is that of a large leguminous tree of Amazon, *Copaiba langsdorfii*, copaiba - a tree of 30 metres high with 1-1.5 m diameter of trunk with a single - cup bore of tapping can yield between 10 and 20 liters of hydrocarbon fluid within two hours (Biswas, 2000). Myers (1983) reports that the fluid resembles with diesel fuel so much so that it can be put straight into the tank of a diesel truck and driven away.

In this country exploration for hydrocarbon has come a long way. With the liberalization of economic policy and subsequent introduction of newer and more foreign cars on Indian roads, there is concern not only over the conservation and resource but also on the environment. It is reported that the country derives 38% of its energy requirements from oil and gas. During 1999-2000 the demand of crude oil was projected to be 96.44 MMT from 54.10 MMT in 1989-90; it is expected to reach 175 MMT in 2006-2007. A gap between demand and supply has appeared due to rapid rise in consumption (*Hindustan Times* dated March 31st 2000, *Save foreign exchange - Join the Oil Conservation Drive*).

The species of *Pittosporum* and *Copaiba* are well represented in various parts of the globe (Brazil, West Indies, C & W Africa) and required to be evaluated for hydrocarbon content with sufficient consideration to their biodiveristy and genetic resource. Indian flora is represented by 11 sps. of *Pittosporum*.

Among the potential Indian species having property similar to the above mentioned to certain extent are *Kingiodendron pinnatum, Hardwickia binata, Canarium* species, *Jatropha curcas* (cultivated and naturalized), *Mesua nagesserium* (syn. *M. ferrea*) and its site and sub specific varieties, *Shorea* species, and dendroid euphorbias bearing latex have potential for exploitation for the purpose. Bennet and Jain (1988) reported the distribution of 16 dendroid Euphorbia in India, namely *Euphorbia antiquorum* Linn., *E. antisyphilitica* Zucc., *E. atoto* Forst.f., *E. barnhartii* Croizat, *E. caducifolia* Haines, *E. cattimandoo* Elliot, *E. epiphylloides* Kurz., *E. milii* Desmoul., *E. neriifolia* Linn., *E. nivulia* Buch.-Ham. *E. pulcherrima* Willd. Ex Klotzsch, *E. royleana* Boiss., *E. santapaui* Henry, *E. tirucalli* Linn. *E. tortilis* Rottl. ex Wight. In recent years, discoveries on the occurrence of species diversity of this taxon has been made, for example, *Euphorbia susan - holmesiae* Binoj. & Gopalan (South India) differing from *E. royleana* in bearing 4-winged branches, orbicular leaves (1 cm long), involucre 3-4 mm in diameter and capsules being 3-4 mm in diameter, *E. vajravelui* Binoj. & Balak., differing from *E. antiquorum* by its fleshy, flat and long branches. *E. santapaui* Henry, endemic to south covered top of Mount Agasthya, Kerala - Tamil Nadu border; *E. lactea* Roxb., a species typical of Moluccas introduced in Indian Botanic Garden, Howrah, *E. mayurnathani* Croiz. - a medium sized tree with mammilate podaria, branches not winged, the plant is believed to be extinct from its original habitat i.e. Mount Pallasana, Kerala along Palghat Gap in Western Ghats

(Singh, 1994). For this group of plants, conclusively it could be mentioned that the potential of several species has not been fully tapped for the hydrocarbon content besides the density of occurrence and quantum of biomass available in different parts of the country is not known.

Forest Soils and Microfauna and Flora

There are nearly 30 different kinds of major forest soil types. These types rich in diversity of soil flora and fauna have evolved for thousand years of environmental interaction with earth and life forms. The forest soils contain mycorrhizzae, microfungi, fauna and bacteria. A mg of soil may contain over 10,000 bacteria. The latter may be found useful for variety of purposes such as in food industry, medicine, transgenic studies etc. Nearly 25 Indian soil microorganisms have been patented by USA. These meet the purpose of manufacturing antibiotic compounds (Mitocromin, Capriomycin, Sorbistin, Virginamycin etc.) and antifungal agents.

Forest soil plays a very significant role in carbon sequestration. The latter depends upon the phenological cycles of plant species and the concentration of atmospheric carbon. There is now International Initiative on Soil Biodiversity under specific programme of Convention of Biological Diversity (CBD).

Integration of Indigenous Knowledge on Potential and Lesser-Known Species

Indigenous knowledge on species of secondary importance in environmentally sensitive areas is little and less known. Several of the species of such areas have developed certain modification in their morphological structure and adapted to the nature's vagaries. The identification, functioning, usage of several of such components are well known to the indigenous communities dependent on them. Besides, the knowledge about tending the river, sustainable utilization of land resource, preservation of flora and fauna, development and maintenance of ecological important sites such as sacred grove, preservation plots and other nature reserves, etc. help in developing management strategies for the conservation of biodiversity and genetic resource in present millennium.

Biswas (1998) and Biswas *et al.* (2000a) identified certain specific sites in Himalayan and north eastern region of India that are significant from floristic and ethnobiological information resource point of view. Certain plant groups such as bamboos, rattans and palms, and species of food and nutrition, medicinal value, etc. received greater attention of science and technology during the last millennium, and the indigenous knowledge of such components are required to be integrated and worked out for the best line of action while considering the different environmental priorities of the present millennium.

Need and Priorities for Conservation of under explored and under utilized species

- Inventorization and characterization of species of economic and environmental relevance.
- Selection and collection of multipurpose-cum-multisite species and their germ plasm of environment and conservation significance. For example, species may be pollutants checking, species for biological control, phytopetroleum, biopesticidal, plant - based derivatives for birth control and species for impoverished soil.
- Ecological-cum-economic e valuation of flora and fauna of undergrowth, second and middle storey of forests in terms of environmental attribution and conservation.
- Application and integration of existing knowledge on well-known species to the secondary important species (genetic improvement and interbreeding, introduction, conservation methodology).
- Input of science and technology for deciphering, applying and integrating the indigenous knowledge existing on little and less-known species with respect to different ecosystems.
- Development of database for rapid retrieval of information resource.
- Identification and prioritization of species-specific-cum-environment related components. Emphasis may be laid on species indicative of potential environment of degrading and restoring areas.
- Setting conservation priorities with emphasis on aspects such as rehabilitation/restoration, relocation, *in situ* and *ex situ* measures.

- Quantification, categorization and prioritization of lesser-known species with the point of view of conservation. This should cover ecological assessment, economic evaluation, threat value assessment, and regional/national/international significance.
- Introduction of less-known rare and threatened species in key areas of conservation (Botanic Garden, Arboreta, Field Gene Banks, Germplasm Banks etc.).
- Maintaining reservoir or pool of species and genetic diversity of lesser-known taxa for long time conservation.
- Restocking/introduction of an admixture of lesser-known species of potential value with well known species of timber value of past millennium.
- Multiplication of giant trees or noble hardwoods seasoned with long history of climate change and record of phenological viability.
- Developing inter-disciplinary and inter-institutional networking and collaboration on studies being carried out on lesser-known species of wider scope and common interest.
- Financial support for working on emergent issues of environmental significance.

Literature Cited

Arora, RK and Pandey, A. (1996): Wild edible plants of India: 279-284. ICAR Publication, New Delhi.

Bennet, SSR and Jain, SS (1988): Distribution of dendroid Euphorbia species in India. Paper presented in Workshop on the "*Utilization of Euphorbia Plants and their latex as an alternative to petroleum based substances*" 5th Feb. 1988. Sriram Institute for Industrial Research, New Delhi pp. 31.

Bennet, SSR; Biswas, Sas.; Chandra, V; Singh, PJ; Chandra S. and R. Dayal (1991): Food from Forest. No. 13 pp. 612, Pl. 158 & Photo 30. Indian Council of Forestry Research & Education, Dehra Dun.

Bernatzky, A. (1978): Tree ecology and preservation. Elsevier Pub., New York.

Bhattacharya, AK (1994): Efficacy of tree species towards gaseous pollutants and its significance in air pollution control by plantation of pollution resistant trees. *Indian Forester* 120 (8): 658-669.

Biswas, Sas. (1998): Developing a floristic model for restoration of threatened biodiversity. *Proceed. National Seminar on Biodiversity Conservation - Challenges & Opportunities*. Oct. 9-10, 1998 FRI: 59-60.

Biswas, Sas. Jain, SS and Sumer Chandra (2000a): Some rare and threatened medicinal plants of Himalayas and NE Region of India and strategies for their conservation. *J. Non Timber Forest Products* 7: (1/2):101-111.

Biswas, Sas. (2000): Lesser known forest species for the environment of new millennium - need and priorities for conservation. *Proceed. National Seminar* on "*Environmental perspectives and management strategies* in the new millennium organized by Centre for Advanced Studies in Forest Soil & Reclamation, FRI and CREATE (Centre for Research on Ecology and Environmental Application, 14-15). April 2001: 20-21.

Biswas, Sas.; Chandra S. and SS Jain (2001): Collecting indigenous knowledge on forest biodiversity conservation and benefit sharing of intellectual property. In: Biswas Sas. (Course Director). Short Term Training Course on Biodiversity Conservation, 7-11th Feb. 2000: 107-112, Forest Research Institute, Dehra Dun.

Biswas, Sas. and TC Bhuyan (1983): On the identity of some food plants of Garo Hills, Meghalaya *Indian J. Forestry* 6 (3): 208-213.

Ghouse, AKM; Khan, FA *et al.* (1984): Effect of air pollution on wood formation in *Tectona grandis*. *Indian J. Forestry* :66-67.

Lanphear, FO (1971): Urban vegetation: Values and stresses. *Hort. Science* 6: 332-334.

Myers, M. (1983): A width of wild species - store house for human welfare. Westview Press, Colarado USA.

Schmid, JA (1975): Urban vegetation. Deptt. of Geography. University of Chicago, USA.

Sharma, YML (1980): Production and utilization of bamboos and related species in the south Asian region in the rural sector. *Indian Forester* 114 (10): 603.

Singh, M. (1994): Succulent Euphorbiaceae of India. pp. 55:A-162 Sector 40 NOIDA-201 303, New Delhi.

Tewari, DN (1992): Monograph on Neem (*Azadirachta indica* A. Juss.) pp. 247. International Book Distributors, Dehra Dun.

Wolverton, BC; Johnson, A. and K. Bounds (1989): Interior landscape plants for indoor air pollution abatement. NASA, Steenis Space Centre, MS.

Chapter III

Forest Ecosystem Diversity of India

Forests are the great reservoir of biodiversity of natural terrestrial ecosystem. Several cultivated and domesticated diversity of plants and animals behind the welfare of the mankind had origin in forest and in its variety of modified landscape. Forest with its diversity due to varied topography, bioclimate, physiognomy and structure, physiography, function and soil is classified into various types. These types are subjected to varying impacting factors. The latter circumscribes the area of occupancy of type. Under the meterorological regime of climate, altitude, latitude, rainfall and temperature various forest types are found in different regions of the country. The temperature influences the forest and classifies it, *sensu lato*, into tropical, sub-tropical, temperate and alpine/arctic. The total annual rainfall received in any part of the country has wide variation. The precipitation between 250-500 cms has resulted into the development of unique vegetation and microhabitats in a particular terrestrial ecosystem. The well crowned/canopied forest types are largely due to the role of solar radiation and the moisture ladden humus rich soil through the phenomenon of evapo-transpiration. The pedological properties of a particular site also results into a specific forest type which spreads with the area under the soil type. A change in the latter is indicatead by a specific composition and dominance of a particular species.

India has 21.6 percent of forest cover to land area compared to 29.6 percent to that of the whole world, 21.8 of Africa and 17.8 of Asia (FAO, 2005). The forest cover of India by and large physiographically occupies the Himalaya, Indo-Ganga Plain, Deccan Plateau, Thar desert and Coastal plain and islands. The different forest types provide niches to 1000 species of birds, 350 species of reptiles, 150 of amphibians, 250 species of fish over 50,000 species of insects, besides 300 species of mammals play an interdependent dynamic role as a part of forest biodiversity and ecosystems. The wild relatives of 4000 species of cultivated crops covering cereals, pulses, millets, fibre yielding species, oilseed, sugarcane, tobacco, spices, condiments, vegetables, fruits and medicinal plants are still found in rich forest types of the country.

Physiognomy, structure, function, floristics, dynamics, environment and habitat, and history of sites with significant forest add to the diversity of the forest types of India. The physiography coupled with edaphic features and types of forest classifies the Indian forest regions as following:

I. Western Himalaya (Jammu & Kashmir, Himachal Pradesh, Uttaranchal and Uttar Pradesh).

(i) Siwaliks (*Shorea robusta* dominant).

(ii) Lower Himalaya (*Pinus roxburghii*) upto 2000 m replaced by *Quercus leucotrichophora, Cedrus deodara* and *Quercus himalayana* above 2000 m.

(iii) Central/main Himalaya. (*Cedrus deodara* and *Pinus wallichiana* above 1700 m followed beyond by *Abies* and *Picea*.

(iv) Trans-Himalaya (Alpine scrub and glacial moraines).

II. Eastern Himalayas [Sikkim, North Bengal (West Bengal) and Arunachal Pradesh].

(i) Sub-montane (*Shorea robusta* dominated and associated with *Michelia, Anthocephalus* etc.). Plains with *Dalbergia sissoo* and *Acacia catechu*.

(ii) Lower Himalaya (Dominance of *Shorea robusta* at lower and species of oaks such as *Quercus lineata, Q. lamellosa* at higher elevation).

(iii) Central Himalaya.

a. Temperate forests (Species of *Abies, Picea* and *Juniperus*).

b. Alpine forests (Cover the limit of timber line with stunted tree growth, meadows and grassy growth.

III. North eastern India (Assam, Manipur, Nagaland, Mizoram, Tripura and Meghalaya).

The forests of this region have the admixture of transitional elements typical of Tibet and Myanmar. The species providing a dense cover belong to *Pterygota alata, Dipterocarpus alata, Shorea robusta, Terminalia myriocarpa, Lagerstroemia regina, Shorea assamica, Endospermum chinense, Mesua ferrea, Pinus kesiya,* bamboo etc. Ecotone of *Pinus kesiya* and *Schima wallichii* in Khasi Hills of Meghalaya is of phytogeographical significance.

IV. Indus Plain (Haryana, Punjab and parts of Rajasthan).

By and large the original forests of this region have been converted into agricultural fields and degraded areas. General landscape is studded with the species of *Acacia nilotica, A. catechu, A. senegal, Balanites roxburghii, Boswellia serrata, Prosopis* sps. (introduced and *P. cineraria*), *Salvadora species, Ziziphus* species etc.

V. Ganga Plain (Uttar pradesh, Parts of Uttaranchal, Bihar, Jharkhand and West Bengal)

(i) Upper Ganga Plain (Eastern Uttar Pradesh with *Acacia catechu Dalbergia sissoo, Tamarix* sps. *Mangifera indica, Saccharum munja S. spontaneum, Erianthus, Imperata* etc.

(ii) Middle Ganga Plain (Parts of eastern Uttar Pradesh and the northern parts of Bihar). The forests of this region are represented by *Acacia catechu, Alstonia scholaris, Anogeissus latifolia, Butea monosperma Madhuca indica, Sterculia urens, Terminalia arjuna, Ziziphus* sps. *Carissa spinarum* other shrubby undergrowth etc.

(iii) Lower Ganga Plain (West Bengal). This comprises of delta of Sundarbans with dominance of species of *Avicennia officinalis Heritiera littoralis, H. minor, Rhizophora mucronata, Bruguiera gymnorrhiza* etc.

(iv) Central India (Parts of Chhatisgarh, Madhya Pradesh, Maharashtra Gujarat, Jharkhand (Bihar) and Orissa. The forests of this region are distributed in the Aravallis, Vindhyas, Satpuras and Chhota Nagpur Plateau (Jharkhand).

The locality factors of this forest region are responsible for the following types of forest.

(i) *Tectona grandis* (Teak) forests.

(ii) *Shorea robusta* (Sal) forests.

(iii) Mixed deciduous forests with *Anogeissus, Diospyros, Terminalia Butea, Dalbergia latifolia, Lagerstroemia* species etc.

(iv) Thorny forests which have come up as a result of intensive biotic influences such as illicit felling, removal of forest produce, forest fire grazing and indiscriminate lopping.

(v) Ecotone of sal (in northern part) and teak in southern part particularly in Bori forests is of phytogeographical significance.

VI. South India (Andhra Pradesh, Tamil Nadu and southern part of Maharashtra encompassing the Deccan plateau).

The main timber and forest produce yielding species of this region are *Tectona grandis, Santalum album, Diospyros melanoxylon, Hardwickia binata, Toona ciliata* etc.

VII. Western Ghats/Malabar Coast (Covers mountainous region of Nilgiris and Pulney Hills of Tamil Nadu and Kerala, Gujarat coast in the north to Kanyakumari in the south).

The region is known for its endemic elements in its mixed deciduous, tropical moist evergreen, sub-tropical or temperate evergreen, semi-evergreen, mangrove and coastal forests.

The principal species are *Dipterocarpus indicus, Kingidendron pinnata, Pterygota alata, Michelia nilgarica, Rhododendron nilgarica, Tectona grandis, Dalbergia latifolia*; among the species of bamboo the taxa of *Pseudoxytenanthera, Ochlandra, Dendrocalamus* etc. are common.

VIII. Bay Islands (Andaman and Nicobar).

The forests of this region are composed of semi-evergreen, evergreen and mangrove forests.

The species are typically rhizophoraceous members, *Calophyllum* sps., *Pterocarpus dalbergioides,* Dipterocarps, Sapotaceous trees, *Terminalia alata, Magnoliaceae, Sterculiaceae,* etc.

Forests of the country have been variously classified. Puri (1960) has classified the forests of India as shown below:

I. Alpine vegetation of India

- Alpine stony deserts
- Alpine scrub
- Alpine meadows or pastures
- Alpine forests
- Alpine fir-birch forest
- Birch-Rhododendron forest.

II. Temperate forests of India

- Himalayan moist temperate forests
- Southern wet temperate forests
- Northern wet temperate forests
- West Himalayan temperate forests

III. Tropical forests of India

- Moist tropical wet evergreen forests
- Southern tropical wet evergreen forests
- Northern tropical wet evergreen forests
- Southern moist tropical semi-evergreen forests
- Northern moist tropical semi-evergreen forests

IV. Tropical moist deciduous forests

- Southern tropical moist deciduous forests
- Teak forests
- Sandal forests
- Northern tropical moist deciduous forests
- Sal forests
- Southern tropical dry deciduous forests
- Northern tropical dry deciduous forests

V. Montane subtropical forests

- Southern sub tropical wet hill forests
- Northern sub tropical hill forests.

VI. Edaphic forests of India

- Moist tropical forests
- Dry tropical forests
- Moist temperate riverain forest
- Mangrove forests of India (Represents *Rhizophorous* members)
- Beach forests
- Tidal forests
- Sunderbans forests

VII. Terrestrial edaphic forest communities of India

- North Indian moist deciduous forest
- North Indian moist *Terminalia* forest
- East Himalayan moist deciduous forest
- East sub Himalayan wet mixed forest
- Wet plains sal forest
- Secondary moist bamboo brakes
- Moist deciduous sal forests

VIII. Biotic forests of India

- Dry evergreen forests
- Southern thorn forests
- Northern thorn forests
- North Indian upper alluvial savannah
- North Indian lower alluvial savannah
- Dry savannah
- Nilgiri sub tropical hill savannah
- Desert vegetation of India
- Vegetation of the Gangetic valley

IX. Grass land vegetation of India

- *Dichanthium-Cenchrus-Erianthus* type
- *Sehima-Dicanthium* type
- *Phragmites-Saccharum* type

X. Bio-edaphic forests of India

- *Anogeissus pendula* forests
- Island dune scrub
- Gangetic saline scrub
- Dry bamboo brakes
- *Hardwickia, Boswellia, Butea* and Babul forests
- Laterite thorn forest
- Artificial forests of India

Forest Types of India (Champion and Seth 1968) are broadly classified into following major groups. These groups are classified according to bioclimatic norms representing storied and architectured forest species ranging from upper storey to the level of undergrowth.

Major types of forests:

I. Moist Tropical Forests

Group:

1. Tropical wet evergreen
2. Tropical semi evergreen
3. Tropical moist deciduous
4. Littoral and swampous

II. Dry Tropical Forests

Group:

5. Tropical dry deciduous
6. Tropical thorn forests
7. Tropical dry evergreen

III. Montane subtropical Forests

Group:

8. Sub tropical broad leaved forests

9. Sub tropical pine forests
10. Sub tropical dry evergreen forests

IV. Montane temperate forests

Group:

11. Montane wet temperate forests.
12. Himalayan moist temperate forests.
13. Himalayan dry temperate forests.

V. Sub-Alpine Forests

Group:

14. Sub-alpine forests

VI. Alpine scrub

Group:

15. Moist alpine scrub
16. Dry alpine scrub

FOREST TYPES

Area under different major forest types indicates that the tropical dry deciduous type covers major portion accounting to 38.2 per cent whereas tropical moist deciduous forests is pre-dominant type forming 30.3 percent of the total forest area of the country.

Champion and Seths' Revised Classification of Indian Forest Types published in 1968 classify the forests of the country mainly into 1. Moist tropical forest, 2. Tropical semievergreen forests, 3. Tropical moist deciduous forests, 4. Littoral and swamps forests, 5. Tropical dry deciduous forests, 6. Tropical thorn forests, 7. Tropical dry evergreen forests, 8. Sub-tropical broad leaved hill forests, 9. Sub tropical pine forests, 10. Sub tropical dry evergreen forests, 11. Montane wet temperate forests, 12. Himalayan moist temperate forests, 13. Himalayan dry temperate forests, 14. Sub alpine forests, 15. Moist alpine scrub, and 16. Dry alpine scrub.

Different Forest Types in India

(A) *Moist Tropical Forest*

Group I : Tropical Wet Evergreen Forests - 8 subtypes
Group II : Tropical Semi-evergreen Forests - 14 subtypes
Group III : Tropical Moist Deciduous Forests - 17 subtypes
Group IV : Tropical SwangsForests - 11 subtypes

(B) *Dry Tropical Forest*

Group V : Tropical Dry Deciduous Forests - 24 subtypes
Group VI : Tropical thorn Forests - 11 subtypes
Group VII : Tropical dry evergreen Forests - 1 subtypes

Montane Subtropical Forests

Group VII : Subtropical broad-leaved Hill Forests - 7subtypes
Group IX : Himalayan mist temperate Forests - 20 subtypes
Group X : Himalayan dry temperate Forests - 8 subtypes

Montane Temperate Forests

Group XI : Montane Wet tempeate forest - 6 subtypes
Group XII : Himalayan moist temperate Forests - 20 subtypes
Group XIII : Himalayan dry temperate Forests - 8 subtypes

Sub alpine

Group XIV : Sub-alpine forest - 5 subtypes

Alpine Forests

Group XV : Moist alpine scrub - 2 subtypes
Group XVI : Dry alpine scrub - 2 subtypes

1. MOIST TROPICAL FORESTS

GROUP 1 – TROPICAL WET EVERGREEN FORESTS

Sub-group 1A Southern tropical wet evergreen forests:

C1 Giant evergreen forests
C2 Andamans tropical evergreen forest
E1 Andamans moist deciduous forests
E3 Southern hilltop tropical evergreen forest
C4 West Coast tropical evergreen forest

Sub-group 1B - Northern tropical wet evergreen forests:

C1 Assam Valley tropical wet evergreen forests (*Dipterocarpus*)
C2 Upper Assam Valley tropical evergreen forest
    (a) *Kayea* forest
    (b) *Mesua* forest
    C3 Cachar tropical evergreen forest

General edaphic seral types of wet evergreen forests:

E1 Cane brakes
E2 Wet bamboo brakes
2S1 Pioneer Euphorbiaceous scrub.

GROUP 2 - TROPICAL SEMI-EVERGREEN FORESTS

Sub-group 2A - Southern tropical semievergreen forests:

C1 Andamans semi-evergreen forests
C2 West Coast semi-evergreen forest
C3 Tirunelveli semi-evergreen forest
2S1 West Coast secondary evergreen Dipterocarp forest

Sub-group 2 B - Northern tropical semi-evergreen forests:

C1 Assam Valley semi-evergreen forest
1a Assam alluvial plains semi-evergreen forest

1b Eastern submontane semi-evergreen forest
1S1 Sub-Himalayan light alluvial semi-evergreen forest
1S2 *Syzygium* parkland
2S1 Pioneer *Euphorbiaceous* scrub
2S2 Eastern alluvial secondary semi-evergreen forest
2S3 Sub-himalayan secondary wet-mixed forest
C2 Cachar semi-evergreen forest
C3 Orissa semi-evergreen forest

General edaphic and seral types of semi-evergreen forests:

E1 Cane brakes
E2 Wet bamboo brakes
E3 Moist bamboo brakes
E4 Laterite semi-evergreen forest
2S1 Secondary moist bamboo brakes

GROUP 3 – TROPICAL MOIST DECIDUOUS FORESTS

Sub-group 3A Andamans moist deciduous forests.

C1 Andamans moist deciduous forest
2S1 Andamans secondary moist deciduous forest

Sub-group 3 B - South Indian moist deciduous forests.

Nearly 5 different types of teak forest are found in this kind.
C1 Moist teak-bearing forest
1a Very moist teak forest
1b Moist teak forest
1c Slightly moist teak forest
C2 Southern moist mixed deciduous forest
2S1 Southern secondary moist mixed deciduous forest.

Sub-group 3 C - North Indian moist deciduous forests.

Diversity of nearly 20 different types of sal forest occur in this particular type. The precipitation has dominant role in governing the biodiversity.

C1 Very moist sal-bearing forest

1a Eastern hill sal forest
- (i) East Himalayan sal
- (ii) Khasi hill sal

1b Eastern bhabar sal forest
- (i) East Himalayan upper bhabar sal
- (ii) East Himalayan lower bhabar sal

1c Eastern terai sal forest
1d Peninsular (coastal) sal forest
C2 Moist sal-bearing forest
- 2a Moist Siwalik sal forest
- 2b Moist bhabar sal forest
- (i) Bhabar-dun sal
- (ii) Damar sal

2c Moist terai sal forest
2d Moist plains sal forest
- (i) Western light alluvium plains sal
- (ii) Chandar sal
- (iii) Eastern heavy alluvium plains sal
- (iv) Kamrup sal

2c Moist peninsular sal forest
- (i) Moist peninsular high level sal
- (ii) Moist peninsular low level sal
- (iii) Moist peninsular valley sal

DSI Moist sal savannah
C3 Moist mixed deciduous forest (without sal)
- 3a West Gangetic moist mixed deciduous forest
- 3b East Himalayan moist mixed deciduous forest.

2S1 Northern secondary moist mixed deciduous forest
2S2 Secondary *Euphorbiaceous* scrub

General edaphic and seral types of moist deciduous forests:

E1 *Terminalia tomentosa* forest
1S1 Low alluvial savannah woodland (*Albizia - Bombax*)
1S2 Eastern hollock forests (*Terminalia myriocarpa*)
  2a *Terminalia - Lagerstroemia*
  2b *Terminalia - Duabanga grandiflora*
2S1 Dry bamboo brakes

GROUP 4 – LITTORAL AND SWAMP FORESTS

Sub-group 4A – Littoral forests:
L1 Littoral forest

Sub-group 4B - Tidal swamp forests:
TS1 Mangrove scrub
TS2 Mangrove forest
TS3 Saltwater mixed forest (*Heritiera littoralis, H. fomes,* and *H. minor*)
TS4 Brackish water mixed forest (*Heritiera* sp. and same as above)
E1 Palm swamp

Sub-group 4C - Tropical freshwater swamp forests:
With the shrinkage of the area under this type aquatic biodiversity is under severe threat.
FS1 *Myristica* swamp forest
FS2 Submontane hill valley swamp forest
FS3 Creeper swamp forest

Sub-group 4D - Tropical seasonal swamp forests:
These forests provide niche to water birds.
SS1 Eastern seasonal swamp forest
SS2 *Barringtonia acutangula* swamp forest
SS3 *Syzygium cumini* swamp low forest
SS4 Eastern seasonal swamp low forest
(*Cephalanthus naucleoides* - Found inside the hilly terrain of North Bengal, Burnihat Valley near Marakdola and Sikkim Himalaya)
SS5 Eastern *Dillenia* swamp forest
2S1 *Syzygium* parkland
2S2 Eastern wet alluvial grassland

Sub-group 4E - Tropical riparian fringing forests:
RS1 Riparian fringing forest

II - DRY TROPICAL FORESTS

GROUP 5 - TROPICAL DRY DECIDUOUS FORESTS

Sub-group 5A - Southern tropical dry deciduous forests:
C1 Dry teak - bearing forest
- 1a. Very dry teak forest
- 1b. Dry teak forest

C2 Dry red sanders (*Pterocarpus santalinus*) - bearing forest
C3 Southern dry mixed deciduous forest

Sub-group 5B - Northern tropical dry deciduous forests:
Microhabitats with sal are quite conspicuous in the type.
C1 Dry sal-bearing forest
- 1a Dry siwalik sal forest
- 1b Dry plains sal forest
- 1c Dry peninsular sal forest

C2 Northern dry mixed deciduous forest

Degradation stages of tropical dry deciduous forests:
DS1 Dry deciduous scrub
DS2 Dry savannah forest
DS3 *Euphorbia* scrub
DS4 Dry grassland

General edaphic types of dry deciduous forests:
E1 *Anogeissus pendula* forest
DS2 *Anogeissus pendula* scrub
E2 *Boswellia serrata* forest
F3 Babul (*Acacia nilotica*) forest
E4 *Hardwickia* forest
F5 *Butea monosperma* forest
F6 *Aegle marmelos* forest
F7 Laterite thorn forest
E8 Saline/alkaline scrub savannah
8a *Phoenix* savannah

8b Babul *(Acacia nilotica)* savannah
8c *Salvadora oleoides - Tamarix dioica* scrub
E9 Dry bamboo brakes

General seral types of dry deciduous forests:
1S1 Dry tropical riverain forest
1S2 Khair - sissoo (*Acacia catechu - Dalbergia sissoo*) forest
1S3 Inundation babul forest
2S1 Secondary dry deciduous forest

GROUP 6 - TROPICAL THORN FORESTS

Sub-group 6A - Southern tropical thorn forests:
C1 Southern thorn forest
C2 Karnatak umbrella thorn forest
DS1 Southern thorn scrub
DS2 Southern *Euphorbia* scrub

Sub-group 6B - Northern tropical thorn forests:
C1 Desert thorn forest
C2 Ravine thorn forest
DS1 *Ziziphus mauritiana* var. *fruticosa* scrub
DS2 Tropical *Euphorbia* scrub

General edaphic, degraded and seral types of thorn forests:
E1 Euphorbia scrub
E2 *Acacia senegal* forest
E3 Rann Saline thorn forest
E4 *Salvadora oleoides* scrub
DS1 *Cassia auriculata* scrub typically found in southern India.
1S1 Desert dune scrub

GROUP 7 - TROPICAL DRY EVERGREEN FORESTS
C1 Tropical dry evergreen forest
DS1 Tropical dry evergreen scrub

III - MONTANE SUBTROPICAL FORESTS

GROUP 8 - SUBTROPICAL BROADLEAVED HILL FORESTS

Sub-group 8A - Southern subtropical broad -leaved hill forests:
C1 Nilgiri subtropical hill forest. Remnant of this type found as *shola*

forest.
DS1 South Indian Sub-tropical hill Savannah (woodland).
E1 Reed brakes (*Ochlandra* spp. - several are endemic.)
C2 Western subtropical hill forest
DS Degradation stages of Southern subtropical broad-leaved hill forests

Sub-group 8 B - Northern subtropical broad-leaved wet hill forest
C2 Khasi subtropical wet hill forest
2S1 Assam subtropical pine forest - This is principally represented in Shillong plateau. The species has remarkable ecotone with *Schima wallichii* in south of Meghalaya.
DS1 Assam subtropical hill savannah woodland

GROUP 9 - SUBTROPICAL PINE FORESTS
C1 Himalayan subtropical pine (*Pinus roxburghii*) forest
1a Lower or Siwalik chir pine (*Pinus roxburghii*) forest
1b Upper or Himalayan chir pine forest
DS1 Himalayan subtropical scrub
DS2 Subtropical *Euphorbia* scrub
C2 Assam (Meghalaya) subtropical pine forests
DS1 Assam subtropical pine Savannah

GROUP 10 - SUBTROPICAL DRY EVERGREEN FORESTS
C1 Subtropical dry evergreen forest
1a *Olea cuspidata* scrub forest
1b *Acacia modesta* scrub forest
DS1 *Dodonea viscosa* scrub

IV - MONTANE TEMPERATE FORESTS

GROUP 11 - MONTANE WET TEMPERATE FORESTS
Sub-group 11A - Southern montane wet temperate forests:
C1 Southern montane wet temperate forest.
DS1 Southern montane wet scrub.
DS2 Southern montane wet grassland.

Sub-group 11A - Southern montane wet temperate forests:
C1 Southern montane wet temperate forest
DS1Southern montane wet scrub

DS2 Southern montane wet grassland

Sub-group 11B - Northern montane wet temperate forests:

C1 East Himalayan wet temperate forests

1a *Lauraceous* forest (Dominate with the species of *Cinnamomum, Persea, Phoebe* etc.).

1b Ban oak forest (*Quercus leucotrichophora*)

1c High-level oak forest

C2 Naga hills (Nagaland) wet temperate forests

GROUP 12 - HIMALAYAN MOIST TEMPERATE FORESTS

C1 Lower Western Himalayan temperate forests

1a Ban oak forest (*Quercus leucotrichophora*)

1b Moru oak forest (*Quercus himalayana* syn. *Q. dilatata*)

DS1 Oak scrub

1c Moist deodar forest (*Cedrus deodara*)

1d Western mixed coniferous forest (spruce, blue pine, silver fir - *Picea, Pinus wallichiana, Abies pindrow*).

1e Moist temperate deciduous forest

1f Low - level blue pine forest (*P. wallichiana*)

DS1 Oak scrub

DS2 Himalayan temperate secondary scrub

C2 Upper West Himalayan temperate forest

2a Kharsu oak forest (*Q. semecarpifolia*)

2b West Himalayan upper oak/fir forest

2c Moist temperate deciduous forest

C3 East Himalayan moist temperate forest

3a East Himalayan mixed coniferous forest

3b *Abies delavayi* forest (Fir typical of eastern Himalaya)

Degradation stages of Himalayan moist temperate forests:

DS1 Montane bamboo brakes

DS2 Himalayan temperate parkland

DS3 Himalayan temperate pastures

General edaphic and seral types of Himalayan moist temperate forests:

E1 Cypress forest (*Cupressus torulosa*)

1S1 Alder forest (*Alnus nepalensis*)

1S2 Riverain blue pine forest

2S1 Low-level blue pine forest

GROUP 13 - HIMALAYAN DRY TEMPERATE FORESTS

(i) Western types

C1 Dry broad-leaved and coniferous forest (*Quercus ilex - P. gerardiana*)

C2 Dry temperate coniferous forest

2a Neoza pine forest (*P. gerardiana*)

2b Dry deodar forest (*Cedrus deodara*)

DS1 Pohu scrub

DS2 Dry temperate scrub

C3 West Himalayan dry temperate deciduous forest

C4 West Himalayan high-level dry blue pine forest (*P. wallichiana*).

C5 West Himalayan dry juniper forest (*Juniperus macropoda*)

(ii) Eastern types

C6 East Himalalayan dry temperate coniferous forest

E1 Larch forest (*Larix griffithiana*)

C7 East Himalayan dry juniper/birch forest (*Juniperus wallichiana*).

General seral types of dry temperate forests:

1S1 *Hippophae/Myricaria* scrub (*Hippophae salicifolia*)

1S2 *Populus/Salix* forest (*Populus ciliata*)

1S3 Western high-level dry blue pine forest

V- SUB-ALPINE FORESTS

GROUP 14 - SUB-ALPINE FORESTS

C1 West Himalayan sub-alpine birch/fir forest (*Betula utilis/ Abies spectabilis*)

1a West Himalayan sub-alpine fir forest

1b West Himalayan sub-alpine birch/fir forest

C2 East Himalayan sub-alpine birch/fir forest (*Betula* spp./*Abies densa*)

Seral and degraded types of sub-alpine forests:

1S1 Hippophae/Myricaria brakes.

1S2 Deciduous sub-alpine scrub.

2S1 Sub-alpine blue pine (*Pinus wallichiana* forest)

DS1 Sub-alpine pastures

VI - ALPINE SCRUB

GROUP 15 - MOIST ALPINE SCRUB

C1 Birch/Rhododendron scrub forest

C2 Deciduous alpine scrub
E1 Dwarf Rhododendron scrub
E2 Dwarf juniper scrub - *Juniperous* spp.
C3 Alpine pastures

GROUP 16 - DRY ALPINE SCRUB
C1 Dry alpine scrub
E1 Dwarf juniper scrub

It is found that tropical deciduous, both moist and dry forest types occupy nearly ten times larger areas than the other types. These types harbour a very biological diversity and play a very vital role in carbon sequestration. Most of the fauna inhabits in these highly vulnerable types due to various impacting pressures. The various states of the country are rich in types of forest ranging from major types to sub-types, for example, Northeastern states have 50 and Uttar Pradesh including Uttaranchal with 68 types while Andaman and Nicobar Islands are endowed with 11 major types, mangrove and coastal types.

AREA UNDER DIFFERENT FOREST TYPES IN INDIA

S.N.	Forest Type	Area (m ha)	% of the total
1.	Tropical wet evergreen forests (TWE)	4.5	5.8
2.	Tropical semi evergreen forests (TSE)	1.9	2.5
3.	Tropical moist deciduous forests (TMD)	23.3	30.3
4.	Littoral and swamp forests (L & SF)	0.7	0.9
5.	Tropical dry deciduous forests (TDF)	29.4	38.2
6.	Tropical thorn forests (TTF)	5.2	6.7
7.	Tropical dry evergreen forests (TDEF)	0.1	0.1
8.	Sub-tropical broad-leaved Hill forests (S-TBLH)	0.3	0.4
9.	Sub-tropical pine (S-TP)	3.7	5.0
10.	Sub tropical dry evergreen (S-TDE)	0.2	0.2
11.	Montane wet temperate (MWT)	1.6	2.0
12.	Himalayan moist temperate (HMTTF)	2.6	3.4
13.	Himalayan dry temperate (HDTF)	0.2	0.2
14.	Sub-alpine (S-Al)	3.3	4.3
15.	Moist alpine (M-Al)	3.3	4.3
16.	Dry alpine (D-Al)	3.3	4.3

Source: Forest Survey of India (FSI) 1996

Forest Cover of India (Statewise and as per percentage)

State/ UT	Dense Forest (Sq.km)	Open Forest (Sq.km)	Mangrove (Sq.km)	Total Forest cover (Sq.km)	% of Geographic Area	Scrub (Sq.km)
Andhra Pradesh	24,190	19,642	397	44,229	16.08	9,559
Arunachal Pradesh	57,756	11,091	0	68,847	82.21	104
Assam	14,517	9,171	0	23,688	30.20	324
Bihar incl. Jharkhand	13,274	13,200	0	26,474	15.23	1,914
Delhi	35	53	0	88	5.93	3
Goa	995	251	5	1,251	33.79	16
Gujarat	6,430	5,504	1,031	12,965	6.61	2,948
Haryana	449	515	0	964	2.18	191
Himachal Pradesh incl. Chhatisgarh	9,120	3,962	0	13,082	23.50	566
Jammu & Kashmir	11,019	9,422	0	20,441	9.20	3,089
Karnataka	24,832	7,632	3	32,467	16.93	4,489
Kerala	8,429	1,894	0	10,323	26.56	91
Madhya Pradesh	81,619	50,211	0	131,830	29.73	3,853
Maharashtra	26,613	19,951	108	46,672	15.17	7,160
Manipur	5,936	11,448	0	17,384	77.86	177
Meghalaya	5,925	9,708	0	15,633	69.70	261
Mizoram	3,786	14,552	0	18,338	86.99	125
Nagaland	5,137	9,027	0	14,164	85.43	14
Orissa	26,073	20,745	215	47,033	30.21	5,439
Punjab	517	895	0	1,412	2.80	107
Rajasthan	4,309	9,562	0	13,871	4.05	6,921
Sikkim	2,363	755	0	3,118	43.94	386
Tamil Nadu	8,659	8,398	21	17,078	13.13	2,836
Tripura	2,228	3,517	0	5,745	54.79	38
Uttar Pradesh incl. Uttaranchal	22,902	11,114	0	34,016	11.55	1,177
West Bengal	3,565	2,672	2,125	8,362	9.42	98
Andaman & Nicobar Islands	6,515	125	966	7,606	92.21	0
Chandigarh	6	1	0	7	6.14	0
Dadra & Nagar Haveli	159	43	0	202	41.14	10
Daman & Diu	0	3	0	3	2.68	0
Lakshdweep	0	0	0	0	0	0
Pondicherry	0	0	0	0	0	0
Total	377,358	255,064	4,871	637,293	19.39	51,896

Source: FSI (2000)

STATEWISE FOREST TYPE DIVERSITY

Andaman & Nicobar Islands (10 types)

Andamans moist deciduous forest
Andamans semi-evergreen forest
Andamans tropical evergreen forest
Cane brakes
Giant evergreen forest
Littoral forest
Southern hilltop tropical evergreen forest
Submontane hill valley swamp forest.
Tidal swamp forest
Wet bamboo brakes

Andhra Pradesh (25 types)

Boswellia forest
Cassia auriculata scrub
Dry bamboo brake
Dry deciduous scrub
Dry grassland
Dry peninsular sal forest
Dry red sanders-bearing forest
Dry savannah forest
Dry teak forest
Euphorbia scrub
Hardwickia forest
Laterite thorn forest
Littoral forest
Moist peninsular sal
Secondary dry deciduous forest
Slightly moist and moist teak forest
Southern dry mixed deciduous forest
Southern dry tropical riverain forest
Southern *Euphorbia* scrub
Southern moist mixed deciduous forest
Southern secondary moist mixed deciduous forest
Southern thorn scrub
Tidal swamp forest
Tropical dry evergreen forest
Tropical dry evergreen scrub

North East India (Sikkim, Assam, Arunachal Pradesh, Nagaland, Meghalaya, Manipur, Mizoram, Tripura) (47 types)

Alder forest
Assam sub-tropical pine forest
Assam sub-tropical pine savannah
Assam valley semi-evergreen forest
Assam valley tropical wet evergreen forest
Birch/Rhododendron scrub
Cachar semi-evergreen forest
Cacher tropical evergreen forest
Cane brakes
Creeper swamp forest
Cypress forest
Deciduous alpine scrub
Dry alpine scrub
Dwarf juniper scrub
Dwarf Rhododendron scrub
East Himalayan dry juniper/birch forest
East Himalayan dry temperate coniferous forest
East Himalayan moist mixed deciduous forest
East Himalayan moist temperate forest
East Himalayan sub-alpine birch/fir forest
East Himalayan wet temperate forest
Eastern hollock forest
Eastern wet alluvial grassland
Hippophae/Myricaria scrub
Khair-sissu forest
Khasi sub-tropical hill savannah woodland
Larch forest
Low alluvial savannah woodland
Low level blue-pine forest
Moist bamboo brakes
Moist plains sal forest
Moist sal savannah
Montane bamboo brakes
Naga hills wet temperate forest
Northern secondary moist mixed deciduous forest
Pioneer Euphorbiaceous scrub
Riparian fringing forest
Secondary Euphorbiaceous scrub
Secondary moist bamboo brakes
Sub-alpine blue pine forest

Sub-alpine pasture
Sub-Himalayan secondary wet mixed forest
Sub-montane hill valley swamp forest
Tropical seasonal swamp forest
Upper Assam valley tropical evergreen forest
Very moist sal forest
Wet bamboo brakes

Bihar including **Jharkhand** (30 types)

Aegle forest
Anogeissus pendula forest
Boswellia forest
Butea forest
Central Indian subtropical hill forest
Dry bamboo brake
Dry deciduous scrub
Dry grassland
Dry peninsular sal forest
Dry savannah forest
Euphorbia scrub
Inundation babul forest
Khair-sissu forest
Low alluvial savannah woodland (*Bombax - Albizia*)
Moist bamboo brakes
Moist sal savannah
Moist Siwalik, bhabar and peninsular sal forest
Northern dry mixed deciduous forest
Northern secondary moist mixed deciduous forest
Ravine thorn forest
Riparian fringing forest
Saline/alkaline scrub savannah
Secondary dry deciduous forest
Southern dry tropical riverain forest
Sub-montane hill valley swamp forest
Syzygium cumini swamp low forest
Terminalia tomentosa Forest
Tropical *Euphorbia* scrub
West Gangetic moist mixed deciduous forest
Ziziphus scrub

Gujarat (26 types)

Acacia senegal Forest
Anogeissus pendula forest
Boswellia forest
Butea forest
Cassia auriculata scrub
Desert dune scrub
Desert thorn forest
Dry bamboo brake
Dry deciduous scrub
Dry grassland
Dry savannah forest
Dry teak forest
Euphorbia scrub
Littoral forest
Mangrove scrub and mangrove forest
Rann saline thorn forest
Saline-alkaline scrub savannah
Salvadora scrub
Secondary dry deciduous forest
Slightly moist and moist teak forest
Southern dry mixed deciduous forest
Southern dry tropical riverain forest
Southern moist mixed deciduous forest
Southern secondary moist mixed deciduous forest
Tropical *Euphorbia* scrub
Ziziphus scrub

Himachal Pradesh (37 types)

Alder forest
Alpine pastures
Birch/Rhododendron scrub
Cypress forest
Deciduous alpine scrub
Dry alpine scrub
Dry bamboo brakes
Dry broadleaved and coniferous forest (*Quercus ilex/P. gerardiana*)
Dry deciduous scrub
Dry siwalik sal forest
Dry temperate scrub
Dwarf juniper scrub.
Dwarf Rhododendron scrub

East Himalayan sub-alpine birch/fir forest
Himalayan sub-tropical scrub
Himalayan temperate parkland
Himalayan temperate pasture
Himalayan temperate secondary scrub
Hippophae/Myricaria brakes
Hippophae/Myricaria scrub
Khair-sissu forest
Lower or Siwalik and upper or Himalayan chir-pine forest
Lower Western Himalayan temperate forest
Moist Siwalik and moist bhabar sal
Montane bamboo brakes
Northern dry mixed deciduous forest
Oak scrub
Olea cuspidata scrub forest
Riverain blue-pine forest
Sub-alpine pasture
Sub-montane hill valley swamp forest
sub-tropical Euphorbia scrub
Upper West Himalayan temperate forest
West Himalayan dry juniper forest (*J. macropoda*)
West Himalayan dry temperate deciduous forest
West Himalayan high-level dry blue-pine forest (*P. wallichiana*)
West Himalayan sub-alpine birch/fir forest (*Betula/Abies*)

Jammu and Kashmir (36 types)

Alder forest
Alpine pastures
Birch/Rhododendron scrub forest
Cypress forest
Deciduous alpine scrub
Deciduous sub-alpine scrub
Dodonaea scrub
Dry alpine scrub
Dry bamboo brake
Dry broadleaved and coniferous forest (*Quercus ilex/Pinus gerardiana*)
Dry deciduous scrub
Dry temperate coniferous forest
Dry temperate scrub
Dwarf Rhododendron scrub
Himalayan sub-tropical forest
Himalayan sub-tropical pine forest

Himalayan temperate parkland
Himalayan temperate pastures
Himalayan temperate secondary scrub
Hippophae/Myricaria brakes
Hippophae/Myricaria scrub
Lower Western Himalayan temperate forest
Low-level blue-pine forest
Northern dry mixed deciduous forest
Oak scrub
Pohu scrub
Populus/Salix forest
Riverain blue-pine forest
Sub-alpine pasture
Sub-tropical dry evergreen forest
Sub-tropical *Euphorbia* scrub
Temperate deciduous forest
West Himalayan dry juniper forest (*Juniperus macropoda*)
West Himalayan dry temperate deciduous forest
West Himalayan high-level dry blue-pine forest (*Pinus wallichiana*)
West Himalayan sub-alpine birch/fir forest (*Betula/Abies*)

Kerala (25 types)

Cane brakes
Dry bamboo brake
Dry deciduous scrub
Dry savannah forest
Dry teak forest
Laterite thorn forest
Lateritic semi-evergreen forest
Littoral forest
Moist bamboo brake
Moist teak-bearing forest
Myristica swamp and sub-montane hill valley swamp
Nilgiri sub-tropical hill forest
Pioneer Euphorbiaceous scrub
Reed brakes (*Ochlandra*)
Riparian fringing forest
Secondary dry deciduous forest
Southern dry mixed deciduous forest
Southern hill-top tropical evergreen forest
Southern moist mixed deciduous forest
Southern montane wet grassland

Southern montane wet temperate forest
Southern secondary moist mixed deciduous forest
West coast secondary evergreen Dipterocarp forest
West coast tropical evergreen forest
Wet bamboo brakes

Madhya Pradesh including **Chhatisgarh** (38 types)

Anogeissus pendula forest
Anogeissus pendula scrub
Babul forest
Boswellia forest
Butea forest
Cassia auriculata scrub
Central Indian sub-tropical hill forest
Dry bamboo brake
Dry deciduous scrub
Dry grassland
Dry peninsular sal forest
Dry savannah forest
Dry savannah scrub
Euphorbia scrub
Hardwickia forest
Inundation babul forest
Laterite thorn forest
Moist bamboo brake
Moist peninsular sal forest
Moist sal savannah
Northern dry mixed deciduous forest
Northern secondary moist mixed deciduous forest
Ravine thorn forest
Riparian fringing forest
Saline/alkaline scrub savannah
Salvadora scrub
Secondary dry deciduous forest
Slightly moist and moist teak forest
Southern dry mixed deciduous forest
Southern dry tropical riverain forest
Southern moist mixed deciduous forest
Southern secondary moist mixed deciduous forest
Southern thorn forest
Southern thorn scrub
Submontane hill valley swamp forest

Terminalia tomentosa forest
Very dry and dry teak forest
Ziziphus scrub

Tamil Nadu (43 types)

Cane brake
Cassia auriculata scrub
Dry bamboo brake
Dry deciduous scrub
Dry grassland
Dry red sanders bearing forest
Dry savannah forest
Dry teak forest
Euphorbia scrub
Hardwickia forest
Karnatak umbrella thorn forest
Laterite thorn forest
Lateritic semi-evergreen forest
Littoral forest
Mangrove scrub and Mangrove forest
Moist bamboo brake
Nilgiri sub-tropical hill forest
Pioneer Euphorbiaceous scrub
Reed brakes (*Ochlandra*)
Riparian fringing forest
Secondary dry deciduous forest
Secondary moist bamboo brakes
Slightly moist and moist teak forest
South Indian tropical hill savannah woodland
Southern dry mixed deciduous forest
Southern dry tropical riverain forest
Southern *Euphorbia* scrub
Southern hill-top tropical evergreen forest
Southern moist mixed deciduous forest
Southern montane wet grassland
Southern montane wet scrub
Southern montane wet temperate forest
Southern secondary moist mixed deciduous forest
Southern thorn forest
Southern thorn scrub
Sub-montane hill valley swamp forest
Tropical dry evergreen forest

West Coast secondary evergreen Dipterocarp forest
West Coast semi-evergreen forest
Wet bamboo brakes
Wet bamboo brakes
Wet coast tropical evergreen forest

Maharashtra (36 types)

Anogeissus pendula forest
Babul forest
Boswellia forest
Butea forest
Cane brakes
Cassia auriculata scrub
Dry bamboo brake
Dry deciduous scrub
Dry grassland
Dry savannah forest
Dry teak-bearing forst
Euphorbia scrub
Hardwickia forest
Inundation babul forest
Laterite thorn forest
Lateritic semi-evergreen forest
Littoral forest
Mangrove scrub and mangrove forest
Moist bamboo brakes
Riparian fringing forest
Saline/alkaline scrub savannah
Salvadora scrub
Secondary dry deciduous forest
Secondary moist bamboo brake
Slightly moist and moist teak forest
Southern dry mixed deciduous forest
Southern dry tropical riverain forest
Southern *Euphorbia* scrub
Southern moist mixed deciduous forest
Southern secondary moist mixed deciduous forest
Southern thorn forest
Southern thorn scrub
Sub-montane hill valley swamp forest
West Coast semi-evergreen forest
Western sub-tropical hill forest

Wet bamboo brakes

Karnataka (31 Types)

Southern hill-top tropical evergreen forest
West Coast tropical evergreen forest
Cane brakes
Wet bamboo brakes
Pioneer Euphorbiaceous scrub
West Coast semi-evergreen forest
West Coast secondary evergreen Dipterocarp forest
Moist bamboo brakes
Lateritic semi-evergreen forest
Secondary moist bamboo brakes
Slightly moist and moist teak forests
Southern moist mixed deciduous forest
Submontane hill valley swamp forest
Riparian fringing forest
Dry teak forest
Southern dry mixed deciduous forest
Dry deciduous scrub
Dry savannah forest
Euphorbia scrub
Dry grassland
Hardwickia forest
Laterite thorn forest
Dry bamboo brake
Southern dry tropical riverain forest
Secondary dry deciduous forest
Southern thorn forest
Southern thorn scrub
Southern *Euphorbia* scrub
Cassia auriculata scrub
Western sub-tropical hill forest

Orissa (32 types)

Aegle forest
Boswellia forest
Cane brakes
Central Indian sub-tropical hill forest
Dry bamboo brake
Dry deciduous scrub
Dry grassland

Dry peninsular sal forest
Dry savannah forest
Dry teak forest
Hardwickia forest
Khair-sissu forest
Laterite thorn forest
Littoral forest
Low alluvial savannah woodland (*Bombax - Albizia*)
Moist bamboo brakes
Moist peninsular sal
Moist sal savannah
Northern dry mixed deciduous forest
Orissa semi-evergreen forest
Palm swamp
Riparian fringing forest
Secondary dry deciduous forest
Secondary moist bamboo brakes
Slightly moist teak forest
Southern dry tropical riverain forest
Southern moist mixed deciduous forest
Southern secondary moist mixed deciduous forest
Sub-montane hill valley swamp forest
Terminalia tomentosa forest
Tidal swamp forest
Very moist peninsular (coastal) sal

Punjab (22 types)
Anogeissus pendula forest (*Anogeissus pendula* scrub)
Butea forest
Desert dune scrub
Desert thorn forest
Dodonaea scrub
Dry bamboo brake
Dry deciduous scrub
Dry grassland
Dry savannah forest
Dry Siwalik sal forest
Euphorbia scrub
Khair-sissu forest
Moist Siwalik sal forest
Montane bamboo brakes
Northern dry mixed deciduous forest

Saline/alkaline scrub savannah
Salvadora scrub
Secondary dry deciduous forest
Sub-tropical dry evergreen forest
Sub-tropical *Euphorbia* scrub
Tropical *Euphorbia* scrub
Ziziphus scrub

Rajasthan (22 types)

Acacia senegal forest
Aegle forest
Anogeissus pendula forest
Anogeissus pendula scrub
Boswellia forest
Butea forest
Cassia auriculata scrub
Central Indian sub-tropical hill forest
Desert dune scrub
Desert thorn forest
Dry bamboo brakes
Dry deciduous scrub
Dry grassland
Dry savannah forest
Dry teak forest
Dry tropical riverain forest
Euphorbia scrub
Northern dry mixed deciduous forest
Salvadora scrub
Secondary dry deciduous forest
Very dry teak forest
Ziziphus scrub

Uttar Pradesh including **Uttaranchal** (68 types)

Aegle forest
Alder forest
Alpine pastures
Anogeissus pendula forest
Anogeissus pendula scrub
Babul forest
Birch/Rhododendron scrub forest
Boswellia forest
Butea forest

Cane brakes
Cypress forest
Deciduous alpine scrub
Deciduous sub-alpine scrub
Desert thorn forest
Dry alpine scrub
Dry bamboo brake
Dry broad leaved and coniferous forest
Dry deciduous scrub
Dry grassland
Dry sal-bearing forest
Dry savannah forest
Dry teak forest
Dry temperate coniferous forest
Dwarf juniper scrub
Dwarf Rhododendron scrub
East Himalayan sub-alpine birch/fir forest
Himalayan sub-tropical pine forest
Himalayan sub-tropical scrub
Himalayan temperate parkland
Himalayan temperate pastures
Himalayan temperate secondary scrub
Hippophae/Myricaria brakes
Hippophae/Myricaria scrub
Inundation babul forest
Khair-sissu forest
Low alluvial savannah woodland (Salmalia - Albizzia)
Lower Western Himalayan temperate forest
Low-level blue-pine forest
Moist bamboo brakes
Moist sal savannah
Moist sal-bearing forest
Montane bamboo brakes
Northern dry mixed deciduous forest
Northern secondary moist mixed deciduous forest
Oak scrub
Olea cuspidata scrub forest
Populus/Salix forest
Ravine thorn forest
Riparian fringing forest
Riverain blue-pine forest
Saline/alkaline scrub savannah

Salvadora scrub
Seasonal swamp forest (*Barringtonia, Syzygium, Cephalanthus*)
Secondary dry deciduous forest
Secondary moist bamboo brakes
Southern dry mixed deciduous forest
Sub-alpine pastures
Sub-montane hill valley swamp forest
Sub-tropical Euphorbia scrub
Terminalia tomentoa forest
Tropical Euphorbia scrub
Upper West Himalayan temperate forest
West Gangetic moist mixed deciduous forest
West Himalayan dry juniper forest (*Juniperus macropoda*)
West Himalayan dry temperate deciduous forest
West Himalayan high-level dry blue pine forest (*Pinus wallichiana*)
West Himalayan sub-alpine birch/fir forest (*Betula/Abies*)
Ziziphus scrub

West Bengal (36 types)

Alder (*Alnus*) forest
Butea forest
Cane brakes
Dry bamboo brake
Dry deciduous scrub
Dry peninsular sal forest
Dry savannah forest
East Himalayan dry temperate coniferous forest
East Himalayan mixed coniferous forest
East Himalayan moist mixed deciduous forest
East Himalayan sub-tropical wet hill forest
East Himalayan wet temperate forest
Himalayan sub-tropical pine forest
Himalayan sub-tropical scrub
Himalayan tempeate pasture
Himalayan temperate parkland
Khair-sissu forest
Larch forest (*Larix griffithiana*)
Littoral forest
Low alluvial savannah woodland (*Bamboo - Albizia*)
Moist sal savannah
Moist sal-bearing forest
Montane bamboo brakes

Northern dry mixed deciduous forest
Northern secondary moist mixed deciduous forest
Palm swamp
Pioneer Euphorbiaceous scrub
Riparian fringing forest
Saline/alkaline scrub savannah
Secondary dry deciduous forest
Secondary moist bamboo brakes
Sub-Himalayan light alluvial semi-evergreen forest
Sub-montane hill valley swamp forest
Terminalia - Duabanga
Tidal swamp forest
Tropical seasonal swamp forest
Very moist hill, bhabar and terai sal forest

Literature Cited:

Champion, HG and Seth, SK (1968): Revised forest types of India. Manager of Publication, Govt. of India, Delhi.

FAO (2005): State of World's Forests. Food and Agriculture Organization of United Nations, Rome.

FSI (Forest Survey of India) (1996): The State of Forest Report - 1995: 1-98, FSI, Dehra Dun.

FSI (Forest Survey of India) (2000): State of Forest Report - 1999: 1-113. Forest Survey of India, Ministry of Environment of Forests, Dehra Dun.

Negi, SS (1989): Forest types of India, Nepal and Bhutan. Periodical Expert Book Agency, Delhi-96.

Puri, GS (1960): Indian forest ecology. Oxford Book Co., Delhi.

Chapter IV

Threatened Forest Flora of India

Introduction

Though late but it has been now realized all over the world that several hundreds of species of flora and fauna are threatened with extinction. It is estimated that nearly 10 per cent of vascular plants fall into the category of threatened species. According to various estimates at least 10% of the living forms in developing countries are threatened. There is tremendous decline of genetic and species diversity and their habitats. FAO (Food & Agriculture Organization) Panel of Experts on Forest Gene Resources in its 5th Session in Rome held in 1979 considered as many as 81 species of forestry importance taxa threatened with extinction or subject to severe genetic depletion in different parts of the world.

There is lacuna in knowledge of ecology, biology, phytogeography and practical conservation of threatened species. Specimens of many species are very poorly and rarely represented in world's herbaria, sometimes as type specimens only. Some of the plants have been collected after a lapse of many decades, some for over a century and many have been eliminated before having recorded. An integrated strategy for the conservation of biodiversity and genetic resources and its efficient and equitable use is yet to be evolved (Swaminathan, 1993). Genetic diversity being one of the related concept of biodiversity covers genetic variability at inter and at intra-specific levels; the latter pertains to the species richness in a particular group of organisms. The functioning of ecological systems depends upon the maintenance of both ecological and genetic diversity.

Forests are the major reservoirs of biodiversity and genetic resources. They harvest solar energy and process mineral resources into usable food, fuel, medicines, fibres, timbers and many other value-added products required by mankind. The forest genetic resources of India are enormous and distributed in different forest types ranging from tropical to alpine types. The area covered by such forest types ranges from 29.4 million hectare of Tropical dry deciduous type with 38.2 per cent and Tropical moist deciduous having 30.3 per cent to 0.2 million hectare of Himalayan dry temperate having 3.4 per cent of the total. As per the Forest Survey of India Report (2000) total forest cover of different States/UTs (Union Territories) is 637,293 Sq.km. with Madhya Pradesh including Chhatisgarh having 131,830 and Union Territory (UT) of Chandigarh 7 sq. km. Specifically major forest types are divided into 16 forest groups and over 200 sub-types according to Champion and Seth (1968). These types are represented in different States of the country.

Forests of Indian region are endowed with rich biodiversity bearing different components for socio-economic transformation of human society. Biodiversity and bioproductivity are interdependent. Each component ranging from physiographic features, geological formations, soil types, floristic and faunistic compositions to nutrient cycling, elements of food, fodder, fibre, fertilizer and fuelwood, medicinal properties of species contributes to the richness of biological diversity and bioproductivity of forest flora (Biswas 1999). It has been rightly referred by Gupta (1997) that the forests and biodiversity are synonymous, the former being the reservoir of latter.

Indian region is considered as the third richest in the world after South America and Malaysia in the number of plant species. There are approximately 45,000 plant species of which *ca* 18000 species are flowering plants. 315 families of these flowering plants are represented in India. As mentioned earlier 60 per cent of the flora are endemic and have so far not been reported anywhere else in the world. The genetic diversity of region such as that of Himalayan flora is unique with important plant groups such as conifers, Rhododendrons, *Quercus* (oaks), *Acer* (maples), *Berberis, Primula, Saxifraga, Saussurea*, orchids, *Hypericum, Impatiens* etc. The Indian floristic region has an admixture of Indo-Malayan, European, Sino-Japanese, Mediterranean and African elements.

Northeastern region of India comprising the States of Arunachal Pradesh, Assam, Mizoram, Manipur, Meghalaya, Nagaland, Sikkim and northern part of West Bengal is the richest zone with nearly 10,000 plant species of flowering plants. For instance, a district in the Gangetic plains has 500-600 species while even a smaller area in the northeast region has over 1000 species. Floristic studies carried out by Biswas *et al.* (1997) in Burnihat Valley, a locality with less than 10 sq. km along the border of Assam and Meghalaya reveals 470 species distributed among 339 genera in 106 plant groups. The region is extremely rich in genetic resources of bamboos, rattans, *Dioscorea, Hedychium, Costus, Michelia*, Orchids, *Gmelina, Shorea, Elaeocarpus Dipterocarpus*, Conifers, medicinal plants etc. (Biswas, 1999).

Indian Region - the Centre of Genetic Diversity

Out of about 2500 species grown world over 320 species have come from Indo-Chinese-Indonesian region, 297 species from Sino-Japanese region, 290 species originated in the African and South American regions. Indian region has contributed about 200 species to the total crop plant genetic resource (Tewari, 1992). The region is a secondary centre for a number of important plants notable among them are the species of bamboo, conifers, *Dalbergia, Acacia, Albizia, Pterocarpus*, oaks, *Rhododendron*, teak, sal, *Dipterocarpus*, mango, betelnut, *Artocarpus*, rattans, *Elaeocarpus, Myrstica*, etc. Out of about 800 species of food plants consumed by the tribals *ca* 300 species occur in India (Biswas 1990, 1991; Tewari, 1992). Among nearly 3000 species of Indian medicinal and aromatic plants 540 find major use as herbal drugs. These species have wide genetic diversity (Arora and Engels, 1993; Biswas *et al.* 1996). Edible food plants have nearly 1500 species from forests and oil yielding seeds of forest origin represent 320 species. Fodder yielding species for useful live-stock are nearly 200.

Endemic Forest Genetic Resource

Restricted distribution of many of the major plant genetic resources has played significant role in evolving several infraspecific variability. The endemic flora of India is very rich and second only to the well-known example of Australia. Over 150 genera with 5000 species are endemic to India. According to Chatterjee (1940) 61.5% of dicots of Indian flora are endemic (Himalaya = 3169 spp; Peninsular India = 2045 spp.). Nearly 5000 plant species (33%) are endemic to Indian region. These are represented by 140 genera (6.5%) with 47 families. Among the largest endemic genera are *Pteracanthus* (20 spp.) and *Nilgirianthus* (20 spp.) of family *Acanthaceae*. The Andaman and Nicobar Islands contribute over 200 species as endemic to the flora of India. Other areas rich in endemism are North-east India, southern parts of Peninsular India and north-west Himalayas. Among the monocots showing high endemism are *Orchidaceae, Pandanaceae, Poaceae, Zingiberaceae, Musaceae, Marantaceae* and *Eriocaulaceae*.

The factors responsible for the high endemism in the biodiversity of Indian region are following: (i) the barrier of Himalayan mountains in the northern part (ii) oceanic system (Bay of Bengal, Indian Ocean and Arabian Sea) (iii) arid condition of western part of India, and (iv) tropical humid conditions of north-eastern and peninsular regions. Not only the endemic elements are urgently required to be protected but also the ecosystem as a whole, i.e. both habitat and community. If a large fraction of above is extinct, it may take millions of years for nature to evolve (Biswas, 1999).

Endemic Forest Species

Some of the important endemic forest species which deserve a mention are *Acacia campbellii* (South India), *Acer hookeri* var. *majus* (E.

Endemically Rich Forest Genetic Resource Areas

Region	Endemic spp.	Species Richness
Peninsular India	2000	6000
Andaman, Nicobar & Lakshadweep Islands	220	2000
NW Himalaya	1000	3500
E. Himalaya	1500	4500
NE India	1000	10,000

Himalaya), *Acer oblongum* var. *microcarpum* (Arunachal Pradesh), *Acer osmastonii* (N. Bengal Himalaya), *A. sikkimense* var. *serrulatum* (Arunachal Pradesh), *Actinodaphnae campanulata* var. *obtusa* (Kerala), *Ailanthus kurzii* (Andaman Is.), *Albizia orissensis* (Orissa), *Allophyllus concanicus* (W. Ghats) *Bentinckia coddapanna* (South India), *B. nicobarica* (Nicobar Is.), *Berberis affinis, B. glaucocarpa, B. lambertii, B. osmastonii* (Uttaranchal), *Bombax insigne* ssp. *anceps* var. *polystemon* (Narcodam Is, Andaman), *Buchanania barberi* (Kerala), *B. platyneura* (Andaman & Nicobar Is.), *Calamus delessertianus* (South India), *C. dilaceratus* (Andaman Is.), *C. inermis* (E. Himalaya), *C. kingianus* (Arunachal Pradesh, Assam), *C. nagbettai* (Karnataka), *Cephalotaxus griffithii* (Arunachal Pradesh), *Crotalaria filipes* var. *trichocarpa* (Western

Area under different forest types

Sl. No.	Forest Type	Area (m ha.)	Percent of the total
1.	Tropical Wet Evergreen forests (TWEF)	4.5	5.8
2.	Tropical Semi-Evergreen forests (TSEF)	1.9	2.5
3.	Tropical Moist Deciduous forests (TMDF)	23.3	30.3
4.	Littoral and Swamp forests (LSF)	0.7	0.9
5.	Tropical Dry Deciduous forests (TDDF)	29.4	38.2
6.	Tropical Thorn forests (TTF)	5.2	6.7
7.	Tropical Dry Evergreen forests (TDEF)	0.1	0.1
8.	Sub-tropical Broad-leaved Hill forests (STBLHF)	0.3	0.4
9.	Sub-tropical Pine forests (STPF)	3.7	5.0
10.	Sub-tropical Dry Evergreen forests (STDEF)	0.2	0.2
11.	Montane Wet Temperate forests (MWTF)	1.6	2.0
12.	Himalayan Moist Temperate forests (HMTF)	2.6	3.4
13.	Himalayan Dry Temperate forests (HDTF)	0.2	0.2
14.	Sub-alpine (S Alp)	3.3	4.3
15.	Moist Alpine scrub (M AlpS)	3.3	4.3
16.	Dry Alpine scrub (D AlpS)	3.3	4.3

Source: FSI Report, 1996

India), *Cryptocarya ferrarsii, Cycas beddomei* (Andhra Pradesh) *Cyanametra beddomei* (W. Ghats), *C. bourdilloni* (W. Ghats-Karnataka), *Dalbergia congesta, D. gardenariana* (Nilgiris, Tamil Nadu), *D. wattii* (Manipur), *Dendrobium tenuicale* (Middle Andaman), *Dipterocarpus kerrii* (Andaman & Nicobar Is.), *Elaeocarpus blascoi, E. gaussenii, E. munronii, E. venustus* (W. Ghats), *Eugenia discifera* (W. Ghats), *Euonymus serratifolius* (Anamalais & Nilgiri hills), *Garcinia cadelliana* (S. Andaman Is.), *Hildegardia populifolia* (SE Ghats), *Hippocratea nicobarica* (Nicobar Is.), *Hopea andamanica* (Andaman & Nicobar Is.), *H. jacobi* (Karnataka Hills), *Humboldtia laurifolia, H. unijuga* (W. Ghats), *Hydnocarpus macrocarpa* (W. Ghats), different species of *Isachne* found in W. Ghats, *Isonandra stocksii* (W. Penninsula), *Kingiodendron*

Major Indian Forest Types (Champion & Seth, 1968)

Sl.No.	Forest Type	Occurrence (State-wise)
1.	Tropical Wet Evergreen Forest.	Ar P, Asm, Mnpr, Ng, Krl, TN, Goa, A & N.
2.	Tropical Semi-Evergreen Forest.	Asm, Ng, Kntk, Guj, Krl, Maha, Orr,TN, A & N, Goa.
3.	Tropical Moist Deciduous Forest.	An P, NE India, Bhr, UP, UA, Guj, Kntk,Maha,Orr,WB,Goa, A & N, D & NH.
4.	Littoral & Swamp Forest.	An P,Orr,TN, Guj, Maha,WB, A & N.
5.	Tropical dry deciduous Forest.	MP, Pb, Raj, UP, UA, Bhr, Har, HP, J & K, Orr, An P, Guj, Kntk, Maha, TN, WB, Krl.
6.	Tropical Thorn Forest.	An P, Kntk, TN, Maha, Guj, Har
7.	Tropical Dry Evergreen Forest.	An P & TN.
8.	Sub-Tropical Broad leaved Hill Forest.	Asm, Maha, Megh, WB.
9.	Sub-Tropical Pine Forest	J&K, UA, HP, Ar P,
10.	Sub-Tropical Evergreen Forest.	J&K, HP, NE India.
11.	Montane Wet Temperate Forest.	Ar P, Mnpr, Skm,
12.	Himalayan Moist Temperate Forest.	J&K, HP, UA
13.	Himalayan Dry Temperate	J&K & HP.
14.	Sub-Alpine Scrub.	J&K, HP, UA, Skm, Ar P.
15.	Moist Alpine	J&K, UA, Skm.
16.	Dry Alpine Scrub	J&K

***Note*:** A & N: Andaman & Nicobar Is.; An P.: Andhra Pradesh; Ar P: Arunachal Pradesh; Asm: Assam; D & NH: Dadar & Nager Havelli; Bhr.: Bihar & Jharkhand, Guj.: Gujarat; HP: Himachal Pradesh; Har.: Haryana; Kntk: Karnataka; Krl.: Kerala; MP: Madhya Pradesh & Chhatisgarh; Maha.: Maharashtra; Mnpr.: Manipur; Ng: Nagaland; Orr.: Orissa; Pb.: Punjab; Raj.: Rajasthan; Skm.: Sikkim; UP: Uttar Pradesh; UA: Uttaranchal; WB: West Bengal; TN: Tamil Nadu.

pinnatum (W. Ghats), *Lagerstroemia hypoleuca* (Andaman Is.), *Livistona jenkinsiana* (NE India), *Madhuca bourdillonii* (SW Ghats), *M. diplostemon, M. insignis* (W. Peninsula), *Magnolia griffithii, M. ustavi* (Assam), *Mahonia jaunsarensis* (Uttaranchal), *Michelia punduana* (NE

India), *Miliusa nilagirica* (W. Ghats), *Nepenthes khasiana* (Meghalaya), *Nilgirianthus circarensis* (Andhra Pradesh, Orissa), *Nothopegia aureo-fulva* (Tamil Nadu Hills), *Ochlandra travancorica* var. *hirsuta* (Tinnevelli Ghats), *Palaquium bourdillonii* (SW Ghats), *Phyllostachys mannii* (E. Himalaya), *Pittosporum eriocarpum* (Uttaranchal Himalaya), *Poeciloneuron pauciflorum* (SW Ghats), *Polyalthia rufescens* (Kerala, Tamil nadu), *Populus gamblei* (E. Himalaya), *Psychotria pendula* (S. Andaman Is.), *Pterocarpus santalinus* (Cuddapah Distt., Andhra Pradesh), *Pyrenaria khasiana* (Meghalaya), *Rhododendron arizelum* (Arunachal Pradesh), *R. formosum* (NE India), *Schizostachyum rogersii* (Andaman Is.), *S. kurzii* (Andaman Is.) *Sinarundinaria longispiculata* (Mizoram), *Sterculia khasiana* (Meghalaya), *Syzygium andamanicum* (Andaman Is.), *S. bourdillonii* (Kerala), *S. palghatense* (Kerala), *Trachycarpus takil* (Kumaon, Uttaranchal), *Uvaria eucincta* (Orissa), *Vateria macrocarpa* (Kerala), *Wendlandia angustifolia* (Tamil Nadu) etc. Endemic elements fall under most threatened categories of plants since these are restricted in distribution and in certain regions under heavy threats. There is urgent need for conservation of endemic elements through *ex situ* and *in situ* measures. Although endemic species can be grown away from its original habitat but the natural processes involved in the development of variabilitiy may become affected and the gene flow confined (Biswas, 1999).

Dynamics of Forest Biodiversity and Genetic Resource Conservation

Integration of genetic conservation concerns into tree improvement and breeding strategies is clearly an absolute and obligate must, and should ensure continued availability of diversity to meet both immediate and future needs related to new social and economic goals; adaptability to changing sites and climates; and survival of the species concerned within the ever-evolving living community of pathogens, pests, competitors and mutualists (Namkoong 1986). Scientific and technical knowledge coupled with financial resources over a span of targeted time ensure improvement, development and effective conservation of forest biodiversity and genetic resources.

The knowledge about the essential components of the ecosystem dynamics is necessary to conserve a wide diversity of species and variation in natural forests. In depth studies on range of complex interactions, for example, pollinators, seed dispersers, litter decomposers, nutrient cycling etc. of natural forests and trees in particular are too meagre. According to Kemp (1992) the diversity found within species viewed as having current or perceived future socio-economic value must be the main target for *in situ* conservation, which should aim at maintaining viable breeding populations and a broad genetic base. Depending on the pattern of distribution of such variation through the stand, as a result of the nature of the species breeding and dispersal systems, highly valuable genetic resources at the intraspecific level may be lost even if the population as a whole survives. A wide selection of provenance from the natural areas of their occurrence is considered to be the safest practicable conservation. The areas for protection of species complex are needed to be given priority. There is need for selective logging operation in mixed tropical forests to maintain a wide genetic diversity of both climax and pioneer species. There has not been much concrete systematic management of fire in complex natural forest communities. The fire drastically reduces species richness. In dry deciduous forest types of Indian region numerous species have adapted pyrophytic nature. According to Norman Myers (1981) and Kemp (1992) there is a major advantage in having a number of conservation land-use categories in the biome, and each nation can design a system of protected areas that meets its own needs, including responsibility for protection of irreplaceable resources and contributing to local development on a sustainable basis.

Forest Flora of India - Centre of Genetic Diversity

Forest Flora of India is rich in genetic diversity widely distributed in coniferous forests of Himalayas with species of *Pinus, Cedrus, Abies, Larix* etc. and broad-leaved forests with species of *Dipterocarpus*, poplars, oaks, *Dalbergia, Acacia, Lauraceae, Tectona* (teak), bamboos, rattans *Michelia, Mesua*, etc. occurring in western Ghats, Northeastern region, eastern Ghats, Gangetic plains and in Andaman and Nicobar Islands at various elevations.

Troup (1921) in his remarkable work on the silviculture of Indian trees emphasizes on over 500 species distributed among 260 genera and over 75 families typical of Indian tree flora. Approximately 300 species of Indian forest flora are enlisted as commercially important timber species (Pearson and Brown, 1932). Himalayan region has 200 timber yielding species (e.g. *Pinus roxburghii, P. wallichiana, Cedrus deodara, Abies species, Picea smithiana, Tsuga dumosa, Larix griffithii, Quercus species, Salix* spp., *Populus ciliata, Alnus nepalensis, Celtis australis* etc.). Tropical forests have significant species diversity among timber yielding genera, for example, *Michelia: M. champaca, M. doltsopa, M. oblonga, M. nilgirica*), *Dipterocarpus (D. macrocarpus, D. alatus, D. turbinatus, D. pilosus*), *Diospyros species, Shorea robusta, S. assamica, S. roxburghii, Albizia (A. procera, A. lucidior, A. chinensis, A. lebbeck), Amoora* species, *Phoebe (P. goalparensis, P. cooperiana, P. hainesiana), Lagerstroemia (L. speciosa, L. parviflora, L. lanceolata), Pterocarpus species, Dalbergia (D. sissoo, D. latifolia, D. procera, D. stipulacea), Acacia (A. nilotica, A. farnesiana, A. sundra*) etc. The Andamans are rich in timber yielding species such as *Bombax insigne, Chukrassia tabularis, Dipterocarpus alatus, D. gracilis, Albizia* spp., *Pterocarpus dalbergioides, Terminalia bialata, T. mannii, Lagerstroemia hypoleuca* (Dogra, 1981). The tropical genera being utilized for various economic values but with less infraspecific variation studied in past are *Tectona grandis, Shorea robusta, S. assamica, Pinus roxburghii, Dalbergia sissoo, D. latifolia, Populus ciliata, Bombax ceiba, Gmelina arborea, Michelia champaca, Dipterocarpus macrocarpus, Altingia excelsa, Mesua nagessarium* (syn. *M. ferrea*), *Vatica lanceaefolia, Pterocarpus dalbergioides, Artocarpus chama, Phoebe goalparensis, Prosopis cineraria, Xylia xylocarpa, Terminalia bialata, Chukrassia tabularis, Cedrus deodara, Dendrocalamus strictus, Bambusa arundinacea, Quercus leucotrichophora, Betula alnoides, Kingiodendron pinnata, Toona ciliata, Celtis australis, Alnus nepalensis, Phoebe goalparensis* etc. Many of the taxonomic forms of taxa mentioned fall under threatened categories (Biswas, 1999).

Several species typical of Indian flora show a wide distribution passing through extreme of climatic and edaphic conditions from sub-Himalayan tracts to the foot hills of eastern and western Ghats. *Rhododendron* species, *Betula* species (*B. utilis, B. alnoides*), *Populus* species, *Acer* species etc. and their wide distribution in Himalayas are not only of phytogeographically significance but also of importance of forest genetic resource of a fragile ecosystem of region. *Magnoliaceae, Lauraceae, Fagaceae, Bambuseae, Dipterocarpaceae, Arecaceae* (rattans and columnar palms), *Myrsticaceae, Moraceae* with their vast range of trees, bamboo and rattans are representatively distributed in rich forests of northeastern region,

Andaman & Nicobar Islands and western Ghats of India. The species have wide spectrum of genetic diversity at infrageneric and intraspecific levels, and several of these areas are under/unexplored.

Natural Variation and Influencing Factors on Forest Flora

With the rise in biotic influence on the whole complex of biodiversity the taxonomic variation among the taxa has become by and large affected. As a consequence little and less known taxa are eliminated before their biological or economic potential could be explored for the sustainable development and utilization. The main factors influencing variability at intrageneric or infra specific levels are the size of species range, amount of environmental diversity within the range, disjunct distribution and factors behind large range species. Variation within taxa evolves geographic races occupying sub-division of the species range, ecological, physiological, chromosomal races and individual genotypes differing from normal ones. Variation associated with geographically different regions in which the species grow results from environmental and genetic factors (Dogra, 1981, 1986). It is observed that in those species and population where the genetic diversity is wide, the species have survived a varying degree of influencing factors.

Major Taxonomic Variability among some Important Forest Tree Species

Among the major Indian forest trees, ecological variation is found in Sal (*Shorea robusta*) in the form of its occurrence in dry and very moist forest types extending from sub-Himalayan tracts in north-western Himalaya (Kangra, Himachal Pradesh) to the plains, and to eastern most part of northeast India (Dijoo Valley). Large scale incidence of mortality among sal trees due to the damage by sal borer (*Hyplocerambyx spinicornis*) in different forest areas of India particularly in Madhya Pradesh (*Hindustan Times*, 12 Dec., 1997) may yield, if worked out through field studies, pest resistant taxonomic variants in the species. Babul (*Acacia nilotica* ssp. *indica* widely distributed in tropical India is well-known by its varieties, namely *teli, vediana* and *cupressiformis* although these may be hithertofore unidentified , infraspecific taxa under the species. *Toona ciliata* - a tropical tree of strong and durable timber is known by three varieties such as *typica, multijuga, pillistaminea* based on foliar and floral characters. Taxonomic studies on genus *Toona* and *T. ciliata* in particular made by Bahadur (1988) reveals 12 varieties, namely vars. *haslettii, listeri, candollei* etc. A wide range of variation in morphological, phenological and geographical features of *Terminalia tomentosa* has resulted into different taxonomic entity of the species. *Bombax ceiba* (semul) shows wide variation in general forms, floral characters and distribution. *Quercus lamellosa, Q. semiserrata, Q. lineata, Q. pachyphylla, Q. spicata*, common oaks of Northeast forests of the country have wide morphological and carpological variations. *Gmelina arborea* is represented in Indian flora by two forms, var. *canescens* and *glaucescens* for variation with grey-pubescent and glabrous undersurface of leaves respectively. *Mesua nagessarium*, a species disjunctly distributed in northeastern region and western Ghat forests having wide range of forms are taxonomically categorised into subspecies, namely ssp. *ferrea, pulchella* and *thewaitesii*. Among the maple, *Acer oblongum* shows two taxonomic forms namely *membranaceum* (NW Himalaya) and *microcarpum* (E. Himalaya), whereas several geographic races among various species exist and there is urgent need to survey and study taxonomically.

Pinus roxburghii, a valuable natural resource of Indian region is differentiated into 101 provenances and races. Based on growth characteristics nine provenances have been recognised by Troup (1921). Phenotypic selection of over 60 superior individuals (plus trees) for better growth and 115 for higher resin yield have been selected from 23 seed stands in Himachal Pradesh (Gupta *et al.* 1970). Taxonomical studies based on leaves have determined variation within species of *Tectona grandis* with regards to bole form and timber characteristics, phyllotaxy, floral biology, biochemical characters, resistance to insect and pests, and phenology. Ecologically teak occurs in five different types experiencing wide range of climatic and edaphic factors. '*Teli*' variety of teak which occurs in Central and Southern India is mainly based on leaf characters, resistance to insects, and phenologically by early leafing and flowering; other infra specific forms are var. *glabrifolia, canescens* and *albudens*. Tewari (1992a) reported 1000 plus trees in teak established in four different National Germplasm Banks. In *Shorea robusta* geographical and ecological variation occur. Strains with characteristically large and small seed in the species have been reported (see Dogra, 1986). Numerous superior forms are identified in *Bombax ceiba* (Arunachal Pradesh 21; Kerala 11; Maharashtra 1 (Anon. 1993). *Acacia nilotica* (babul), a species widely distributed in arid regions of India is taxonomically categorised into 9 sub-species of which five namely, *cupressiformis, indica, adstringens, hemispherica* and *sub-alata* occur as a result of variation and interspecific hybridization in areas of Pakistan and India (Ali and Quiser, 1980). *Cedrus deodara* is (*deodar*) known for its occurrence in different ecological zones of Himalayas such as outer moist, inner dry and inner most arid zones having infraspecific variants. Dogra (1986) mentioned about the morphological and phenological variation in needle, colour and texture of foliage, tree form, growth rate, seed size, germination and disease resistance of *deodar*. Among the bamboos of economic value, ecological and morphological variations occur as following (Biswas, 1995 and Banik, 1995).

Dendrocalamus strictus: Common type of this bamboo occurs in larger areas, ordinary form producing medium-sized hollow to solid culms; Large type - occurs in north India, with out side branches and culms with long, straight and smooth internodes; Dwarf type - Rarely forms culms. Found in parts of Central India. Taxonomic varieties namely, *argentea, prainiana* and *sericeus* (*D. sericeus*) are well known under this species. *Bambusa arundinacea:* On the basis of geographical locations in the Indian sub-continent the bamboo has two main growth forms, the one with large culms commonly found in South India (var. *gigantea*) and other one with thickly branched, spiny, small-culmed found in north India. It is interesting to note here that this species has been mentioned in *Guinness Book of Records* (1990) as an example of tallest grass, reported from Travancore in November 1904 measuring 37 m in height (Biswas, 1999).

Variation with respect to height and girth in forest tree diversity occurs quite remarkably and there are records of *Dipterocarpus macrocarpus* and *Shorea assamica* attaining girths up to 7 m and height upto 50 m. Dogra (1981) mentioned about the maximum dimension attained by some Indian trees e.g., *Abies pindrow* (498 cm girth/62.8 m height), *Adina cordifolia* (782 cm/41.8 m), *Cedrus deodara* (1646 cm/73.2 m), *Dipterocarpus gracilis* (853 cm/47.2 m), *Holoptelea integrifolia* (594 cm/53 m), *Picea smithiana* (752 cm/65.5 m), *Bombax ceiba* (782 cm/41.1 m), *Tectona grandis* (625 cm/58.5 m) and *Terminalia arjuna* (109? cm/30.5 m).

Threatened Forest Genetic Resource of Potential Value

The phytogeographical aspects where two or more taxonomic units of important forest genetic resource occur together or as admixture new infrageneric and/or infraspecific taxa are bound to occur. This is true with the fact that numerous strains and phenotypic expressions among the species at infraspecific levels have been recorded time to time. Some of the important examples are in oaks (*Quercus, Castanopsis, Lithocarpus*), Rhododendrons, *Elaeocarpus, Albizia, Acacia Populus, Amoora, Dysoxylum* etc.

Monotypic groups can be considered as some of the most threatened taxonomic entity since these species do not have chances for developing newer traits through cross hybridizing with other species except at infraspecific levels. Such taxa have limited phylogenetic diversity. Some of the striking examples of such forest genetic resource in Indian floristic region are *Chloroxylon, Indopiptadenia, Indopolysolenia, Kendrickia, Kingiodendron, Ougeinia, Pongamia, Tamarindus, Tecomella, Tetrameles, Tetracentron* etc.

Conservation of Threatened Forest Flora - Status

Status of forestry research advances on the diversity and distribution evaluation on enlisted important and threatened species indicates that the little and less-known elements have not been studied in depth. Recent methods for diversity evaluation such as chemotaxonomical, cytotaxonomical, palynological, morphometric analysis of seedlings and biochemical characterisation such as isozymes, DNA markers, RFLP (Restriction-Fragment-Length-Polymorphisms) and PCR (Polymerase-Chain-Reaction) have not yet been applied. Biswas (1995), Wuehlisch *et al.* (1996) and Biswas and Wuehlisch (1997) have worked out the diversity characterization of *Populus nigra* clones through isozyme studies with a view to widening the gene base. Studies indicate that the taxonomical, ecological and anatomical evaluation of species were by and large restricted to elite species only.

Status of species diversity and research input as worked out in the present treatise shows that there are world over 60 genera representing approximately 5470 forest species including *ca.* 500 typically from Indian region which received higher R & D input. 130 threatened taxa related to the well known species are utilized under different practices.

Newer Trends in Biodiversity Survey and Integration

Evaluation of elite and potential species of forestry and economic importance through floristic surveys, phytogeographical and ecological studies has been carried out by several workers. Application of remote sensing and aerospace data on the threatened habitat, anthropogenic factors in vogue

Species Diversity and Threatened Taxa Research Input

Genera	No.of spp world over	No.of sp p. in India	No.of sp p. with higher input of Research	No.of taxa threatened in India (Exc. site specific examples)
1.	2.	3.	4.	5.
Abies	50	6	1	1
Acacia	*ca*700	25	2	3
Acer	200	*ca* 20	-	6
Ailanthus	10	4	2	1
Albizia	100-150	20	2	7
Anogeissus	11	5	1	1
Anthocephalus	3	1	1	-
Artocarpus	50	18	2	2
Aquilaria	15	2	-	2
Azadirachta	2	1	1	-
Bombax	8	2	1	1
Buchaniana	25	6	1	2
Butea	30	3	1	-
Casuarina	40	3	1	-
Cedrus	4	1	1	-
Celtis	80	8	-	-
Chukrassia	2	2	-	-
Dalbergia	100	27	2	3
Dipterocarpus	75	10	3	1
Elaeocarpus	200	26	-	7
Exbucklandia	2	1	-	-

Contd

1.	2.	3.	4.	5.
Firmiana	8	2	-	1
Gmelina	30	4	1	2
Hardwickia	2	1	-	-
Indopiptadenia	1	1	-	1
Kingiodendron	1	1	-	1
Lagerstroemia	50	10	1	2
Larix	*ca*12	2	-	-
Madhuca	80	4	1	3
Mangifera	40	5	1	3
Mesua	3	1	-	1
Michelia	50	12	1	1
Moringa	10	2	1	1
Ougeinia	1	1	1	1
Palaquium	*ca*120	5	-	1
Phoebe	70	9	1	1
Picea	50	3	1	2
Pinus	100	5	1	2
Pithecellobium	200	5	-	-
Podocarpus	100	2	-	2
Pongamia	1	1	1	-
Populus	35	6	1	2
Prosopis	40	3	1	-
Pterocarpus	100	4	1	1
Pterospermum	40	12	-	2
Quercus	300	20	1	-
Rhododendron	500	80	-	47
Salix	500	35	-	-
Santalum	25	1	1	-
Shorea	180	6	1	-
Syzygium	500	50	1	11
Tamarindus	1	1	1	-
Taxus	10	1	1	1
Tecomella	1	1	-	-
Tectona	3	1	1	-
Terminalia	250	12	2	1
Tetrameles	1	1	-	-
Toona	15	4	-	5
Tsuga	15	1	-	-
Xylia	5	1	-	-
Total (60)	*ca* 5461	*ca* 495	47	129

and range extension of taxa can prove significant for developing conservation strategy on the species. Prance (1984) found that the aerial colour slides at a scale of 1:2000 could be extremely reliable for the identification of over 50% of the tree species in certain plots of the Queensland rain forest of Australia.

Gupta and Biswas (1997) emphasized on the biodiversity characterization of threatened habitats with species of forestry and ethnobiological values using remote sensing (RS) and GIS (Geographic Information Systems) data base. Roy *et al.* (1997) used satellite remote sensing approach (patch size, shape density, interspersion and juxtaposition) to analyze ecosystem using landscape ecological principles in two protected areas, namely Madhav National Park (Madhya Pradesh) and Balphakram National Park (Meghalaya). Nair and Menon (1998) estimated bamboo resources of *Bambusa bambos* in Kerala using remote sensing techniques with a view to developing a rapid assessment of bamboo availability. The studies revealed maximum density class in Muthanga, S. Battery and Kurichiyat ranges respectively with 12.7 sq. km., 7.87 sq. km. and 15.23 sq. km. of area. Prasad *et al.* (1998) made a GIS approach for location of biodiversity hot spots in western Ghats of Kerala. 100 endemic plant species, 16 bird species and 6 mammal species were taken into consideration. Among the significant species were lion - tailed macaque, Nilgiri langur; Malabar grey hornbill, Nilgiri wood pigeon, white - bellied blue fly catcher, great pied hornbill, grey - headed bulbul; *Anaphalis travancorica, Bentinckia condapanna, Calamus travancoricus, Ceropegia* spp., *Cinnamomum travancoricum, Cynometra travancorica, Garcinia indica, Janakia aryapathra, Nilagirianthus lupinus, N. nilgherensis, Podocarpus wallichianus, Rauvolfia beddomei, Vannila wightiana* etc.

There is lacuna in the knowledge of canopy science of forest biodiversity in India. Although canopy walkways are being developed in various parts of forest landscapes of the world but they are meeting presently the aesthetic values and need of ecotourism only. It is suggested by RS and GIS experts that a library of hyperspectral signature of various tree species using the canopy access system, when integrated with high-resolution satellite imageries like ICKNOS, will aid in working out the exact spatial distribution of desired tree species (Devy and Ganesh, 2003; Sanchez - Azofeifa, 2002).

Important Indian Forest Tree Genetic Resource and their Variability

Species	*Infraspecific*	*Phenotypic variants*
Abies pindrow	-	-
Acacia catechu	3	37
A. nilotica	5	44
Acrocarpus fraxinifolius	-	20
Ailanthus excelsa	-	38
Albizia lebbeck	2	-
Albizia procera	-	-
Alnus nepalensis	-	-
Anogeissus latifolia	-	-
Anthocephalus chinensis	-	-
Artocarpus chama	-	-
Azadirachta indica	-	140
Bombax ceiba	-	41
Butea monosperma	-	-
Casuarina equisetifolia	-	171
Cedrus deodara	11	-
Chukrassia tabularis	-	25
Cupressus torulosa	-	-
Dalbergia latifolia	-	-
Dalbergia sissoo	-	300
Dipterocarpus macrocarpus	-	39
Dipterocarpus turbinatus	-	-
Exbucklandia populnea	-	-
Gmelina arborea	2	98
Kingiodendron pinnata	-	-
Lagerstroemia speciosa	2	-
Madhuca latifolia	-	8
Mesua ferrea	3	-
Michelia champaca	-	35
Ougeinia oojeinensis	-	-
Palaquium ellipticum	-	-

Contd

Species	*Infraspecific*	*Phenotypic variants*
Phoebe goalparensis	-	40
Pinus kesiya	-	-
Pinus roxburghii	-	101
Pinus wallichiana	1	-
Pongamia pinnata	-	67
Populus ciliata	-	-
Populus gamblei	-	-
Prosopis cineraria	-	77
Pterocarpus dalbergioides	-	-
Pterocarpus santalinus	-	NA
Quercus leucotrichophora	-	-
Quercus semecarpifolia	-	-
Santalum album	-	92
Shorea assamica	-	35
Shorea robusta	-	50
Syzygium cumini	-	85
Tamarindus indica	-	304
Taxus baccata	14	-
Tectona grandis	3	1000
Tecomella undulata	-	-
Terminalia myriocarpa	-	32
Tetrameles nudiflora	-	-
Toona ciliata	12	-
Xylia xylocarpa	-	-
Sps. (55)	58	2879

Conservation of Threatened Forest Flora

The conservation of wide diversity of forest genetic resource is dependent on the maintenance of essential functional components of the ecosystem and understanding of ecosystem dynamics. According to Kemp (1992) the diversity found within species viewed as having current or perceived future socio-economic value must be the main target for *in situ* conservation, which should aim at maintaining viable breeding populations and a broad genetic base. There is need to demarcate a portion of ecosystem to conserve intraspecies variation or subspecific populations. Many of the specific-species rich forest tracts both in production and conservation forests for example, teak (*Tectona grandis*) forests of Madhya Pradesh and Chhatisgarh, sal (*Shorea robusta*) forests in different bioclimatic regions ranging from northern India (Himachal Pradesh) to Dijoo valley in Assam and Central India, hollock (*Terminalia myriocarpa*), hollong (*Dipterocarpus macrocarpus*), makai sal (*Shorea assamica*), Kadam (*Anthocephalus chinensis*), Khasi pine (*Pinus kesiya*), gamari (*Gmelina arborea*), jarul (*Lagerstroemia regina*), nahor (*Mesua nagessarium)*, bonsum (*Phoebe goalparensis*) forests etc. of northeastern India, sandal (*Santalum album*), red sander (*Pterocarpus santalinus*), Andaman padauk (*Pterocarpus dalbergioides*), gurjan (*Dipterocarpus turbinatus*) forests etc. of southern India and Andamans, chir pine (*Pinus roxburghii*), kail pine (*P. wallichiana*), fir (*Abies* spp.), deodar (*Cedrus deodara*), poplar (*Populus ciliata*), oak forests etc. of Himalayan region are required to be conserved *in situ.* From the natural ranges, strategy should aim at conserving a wide selection of provenances and checking genetic pollution by another provenance of the species (Biswas, 1999). Biswas and Wuehlisch (1997) attempted through isozyme studies to characterise different clones of black poplar (*Populus nigra*) occurring in forest areas of Federal Republic of Germany and adjoining European countries with a view to understanding the genetic impurity among the taxa. Such studies can be made applicable to the Indian taxa.

Gaps in the knowledge of different components of genetic resource, taxonomic and phytogeographical entities, factors responsible for the threatened status of taxa and different aspects of conservation measures can be stored and retrieved through modern computerized systems. The integration of GIS helps in the definition and interpretation of species distribution patterns with respect to various forest types and environmental variables. With the application of GPS (*Global Positioning System)* habitat specific species can be undertaken for inventorization.

The habitats of threatened flora of Indian floristic region including forest genetic resource of economic and potential value published in Red Data Book (Vol. I-III 1987, 1988 & 1990) and categorized as per the IUCN categories of conservation status by Botanical Survey of India have been located on GIS by Roy and Sarnam (1995) of Indian Institute of Remote Sensing, Dehra Dun India.

Conservation through Preservation Plots

Indian region has 89 National Parks (NP) and 489 Wildlife Sanctuaries (WS) representing about 5 per cent of total geographical area. NPs are spread over 1.14 percent (37,530,76 sq. km) and WS over 3.56 percent (154,572,80 sq. km) of geographical area of this country. Preservation of individual species helps in preserving a reservoir of continually evolving genetic material for future use. Bhadran (1958) enlisted the giant trees of different forest species in various forest areas. The species such as that of *Aesculus, Ailanthus, Acacia, Albizia, Betula, Exbucklandia, Gmelina, Mangifera, Mesua, Cedrus, Dalbergia, Dipterocarpus, Michelia, Madhuca, Pinus, Platanus, Shorea, Tamarindus, Tectona, Terminalia, Quercus, Toona* etc. provide sufficient information for gene conservation. Establishing preservation plots and protecting noble trees in them had its statutory beginning in India in the year 1929 after the Third Silvicultural Conference held at Forest Research Institute, Dehra Dun (Khullar 1992). Sal (*Shorea robusta*) forests of Saranda in Bihar constituted in 1905-06 seems to be the oldest preservation plot in India. Similarly quality sal in the Haldwani Division, *Cedrus deodara* and *Picea smithiana* in Chakrata Division of Uttaranchal of erstwhile Uttar Pradesh were established as preservation plots. Rodgers (1991) believes that the plots covering vast range of forest types in India could play a major role in national efforts to protect biodiversity. Preservation plots with whole complex of biodiversity and genetic resources ranging from micro to macro fauna and flora play significant role in conserving, *in situ,* the elements for posterity.

According to Gupta and Totey (1994) there is a lot of variation in the area under preservation plots in India ranging from 0.01 ha in Himachal Pradesh being smallest plot to as large as 4000 ha in Assam. Floristic analysis drawn on the composition of different preservation plots of Madhya Pradesh indicates that the genetic variability at intraspecific levels are well-preserved in species such as *Terminalia tomentosa* (*sensu lato*), *Tectona grandis, Shorea robusta, Boswellia serrata, Hardwickia binata, Bombax ceiba, Madhuca indica, Diospyros melanoxylon, Chloroxylon swietenia,*

Anogeissus latifolia, A. pendula, Butea monosperma, Dalbergia latifolia, Acacia catechu, Sterculia urens, Gmelina arborea etc. In Preservation Plots of Orissa germplasm variability in teak, sal, haldu (*Adina cordifolia*), gamari (*Gmelina arborea*), kadam (*Anthocephalus chinensis*) and bamboos, e.g., *Bambusa arundinacea* and *Dendrocalamus strictus* and their intraspecific variability can be preserved for conservation purposes.

Present Status of Preservation Plots in India

(*Source*: Khullar, 1992; Gupta & Totey 1994; Biswas 1999)

State	No. of Plots		
	Conservation Forests	Production Forests	Total
Andhra Pradesh	11	-	11
Arunachal Pradesh	1	-	1
Assam	9	1	10
Bihar	9	-	9
Gujarat	18	-	18
Himachal Pradesh	6	6	12
Jammu & Kashmir	3	-	3
Karnataka	11	-	11
Kerala	8	43	51
Madhya Pradesh	25	-	25
Maharashtra	11	-	11
Meghalaya	2	-	2
Orissa	6	-	6
Rajasthan	3	-	3
Tamil Nadu	2	67	69
Uttar Pradesh incl. Uttaranchal	32	-	32
West Bengal	27	5	32
Total	184	122	306

Protected areas can play a key role in the conservation and utilization of natural forest genetic resource. The status information on the intraspecific variation among the taxa are inadequate.

Preservation Plots by Forest Types

Type	No. of plots	States/Union Territories
A. **NATURAL FORESTS.**	**Total 163.**	**17**
I **Moist tropical Forests**		
1. Tropical Wet Evergreen	9	Asm, Ar.Pr.,And.,Ktka.
2. Tropical Semievergreen	11	And.,Asm,WB.
3. Tropical Moist Deciduous Forest.	66	And.,Asm.,WB,Ktka, MP,Maha., TN.,Megh., Or.,WB.,UP. & UA.,
4. Littoral and swamp forest	5	UP.
II **Dry Tropical Forest**		
5. Tropical dry deciduous	32	And.Pr,Guj.,HP, Krla, Ktka,Maha.,MP, Raj., UP & UA,WB.
6. Tropical thorn forest	1	Ktka.
7. Tropical dry evergreen forest.	1	TN
III **Montane Subtropical forests**		
8. Subtropical broad leaved hill forest	8	Maha.,WB
9. Sub-tropical pine forest	2	UA
10. Subtropical dry evergreen forest.	-	-
IV **Montane Temperate Forests.**		
11. Montane wet temperate forest	4	WB
12. Himalayan wet temperate forest	17	HP,UA,WB
13. Himalayan dry temperate forest.	-	-
V **Sub-alpine forests**		
14. Subalpine forest	-	-
VI **Alpine Scrub**		
15. Moist alpine scrub	-	-
16. Dry alpine scrub	-	-

Preservation Plots by Forest Types (Species - specific)

	Type		No. of plots	States/Union Territories
	Edaphic types	**Total**	**5**	**4**
	Anogeissus pendula Forest		3	MP,Raj.
	Hardwickia forest		1	And.Pr.
	Cypress forest		1	UA
	Seral types	**Total**	**2**	UP & UA
	Khair-sissoo forest	**Total**	2	UP & UA
	PLANTATION CROPS	**Total**	**25**	**7**
1.	Southern hill top tropical evergreen forest (*Pinus roxburghii*, *P. radiata*)		2	TN
2.	Moist teak bearing forest (Teak, *Swietenia macrophylla*)		2	Krla.
3.	Southern moist mixed deciduous forest (*Swietenia*,Teak, *Ailanthus integrifolia*).		6	Krla,TN
4.	Very moist sal bearing forest (Teak,Sal).		2	Asm.,WB
5.	Southern dry mixed deciduous forest (*Casuarina montana*)		3	TN
6.	Dry sal bearing forest (teak)		1	WB
7.	Southern montane wet grassland (*Acacia mearnsii, Pinus radiata)*.		3	TN
8.	Lower western Himalaya temperate forest (*Larix decidua, Castanea sativa, Fraxinus hookeri*)		6	HP

Abbreviations: And. = Andaman & Nicobar Islands; And. Pr.= Andhra Pradesh; Ar. Pr. = Arunachal Pradesh; Asm. = Assam; Guj. = Gujarat; HP. = Himachal Pradesh; Krla. = Kerala; Ktka. = Karnataka; Maha. = Maharashtra; Megh. – Meghalaya, MP = Madhya Pradesh; Or. = Orissa; Raj. = Rajasthan; TN = Tamil Nadu; UP = Uttar Pradesh including Uttaranchal (UA); WB = West Bengal.

Preservation Plots with some Important Species

Sl. No.	Species	State/Union Territory	No. of Preservation Plots Natural Forests	Plantations	Total
1.	*Tectona grandis*	Andhra Pradesh	2	-	2
		Assam	-	1	1
		Gujarat	13	-	13
		Karnataka	3	-	3
		Kerala	-	2	2
		Madhya Pradesh	8	-	8
		Maharashtra	5	-	5
		Tamil Nadu	1	1	2
		West Bengal	-	1	1
		Total	32	5	37
2.	*Shorea robusta*	Assam	2	-	2
		Bihar	7	-	7
		Madhya Pradesh	8	-	8
		Meghalaya	2	-	2
		Orissa	2	-	2
		Uttar Pradesh incl. Uttaranchal	10	-	10
		West Bengal	7	-	8
		Total	38	1	39
3.	Conifers:				
	Abies pindrow	Uttaranchal	1	-	1
	Abies spectabilis	Himachal Pradesh	2	-	2
	Cedrus deodara	Uttaranchal	4	-	4
	Picea smithiana	Himachal Pradesh	2	-	2
	Pinus roxburghii	Tamil Nadu	-	1	1
		Uttaranchal	2	-	2
	Pinus wallichiana	Himachal Pradesh	1	-	1
		Total	12	1	13
4.	*Dalbergia latifolia*	Andhra Pradesh	1	-	1
		Gujarat	1	-	1
		Karnataka	4	-	4
		Maharashtra	2	-	2
		Total	8	-	8
5.	*Pterocarpus dalbergioides*	Andaman & Nicobar	1	-	1
6.	*Pterocarpus santalinus*	Andhra Pradesh	2	-	2

contd.

Sl. No.	Species	State/Union Territory	No. of Preservation Plots Natural Forests	Plantations	Total
7	*Dipterocarpus macrocarpus*	Arunachal Pradesh	1	-	1
		& Assam	1	-	1
		Total	2	-	2
8.	*Dipterocarpus alatus*	Andaman & Nicobar	2	-	2
9.	*Exbucklandia populnea*	West Bengal	1	-	1
10.	*Phoebe goalparensis*	Assam	3	-	3
11.	*Michelia champaca*	Assam	1	-	1
		West Bengal	1	-	1
		Total	8	-	8
12.	*Santalum album*	Kerala	1	-	1
13.	*Tetrameles nudiflora*	Andaman & Nicobar	1	-	1
14.	*Gmelina arborea*	Meghalaya	2	-	2
15.	*Quercus semecarpifolia*	Uttaranchal	2	-	2
16.	*Bambusa arundinacea*	Karnataka	1	-	1
		Orissa	1	-	1
		Total	8	-	8
17.	*Dendrocalamus strictus*	Uttar Pradesh incl. Uttaranchal	1	-	1

HP = Himachal Pradesh; MP = Includes Chattisgarh; NP = National Park; WS = Wildlife Sanctuary; J & K = Jammu & Kashmir; A & N = Andaman & Nicobar; Bihar = Includes Jharkhand; TN = Tamil Nadu; UP = Uttar Pradesh & includes Uttaranchal; WB = West Bengal.

Priority areas of significance to conservation of India's Biological Diversity & Genetic Resources

(*Based on* Gadgil and Meher-Homji 1986)

Sl.No.	Vegetation Type	Total area under nature reserves (sq.km.)	Key areas for conservation	State
1.	2.	3.	4.	5.
1.	*Calligonum series + Prosopis, Salvadora - Capparis-Ziziphus series* (Thorny Types)	3084.55	Jaisalmer Desert NP	Rajasthan
2.	-do-	-	-	-
3.	*Acacia - Capparis series* (Thorny Types)	75.30	Velavadar NP Bhavanagar Distt.	Gujarat
4.	*Acacia senegal - Anogeissus pendula series* (Deciduous Type)	8.20	-	-
5.	*Acacia catechu - Anogeissus pendula series* (Deciduous Type)	2317.46	Sariska W.S. Alwar Distt. Chambal W.S. Kota Distt. National Chambal S. Etawah - Agra Distt. Sawai Madhopur Distt. Keola Deo Ghana NP, Bharatpur Distt.	Rajasthan Rajasthan Uttar Pradesh Rajasthan Rajasthan
6.	*Anogeissus - pendula - A. latifolia series* (Deciduous Type)	925.16	Shivpuri NP	MP
7.	*Acacia - Anogeissus latifolia series* (Deciduous Type)	-	-	-
8.	*Anogeissus latifolia - Hardwickia binata series* (Deciduous Type)	3901.73	Nagarjunsagar WS, Hydrabad Distt. Tungabhadra WS	Andhra Pradesh Karnataka

1.	2.	3.	4.	5.
9.	*Anogeissus latifolia - Terminalia series* (Deciduous Type)	954.65	-	-
10.	*Terminalia - Anogeissus latifolia - Cleistanthus series* (Deciduous Type)	1079.71	Nagzira WS Bhandara Distt.	Maharashtra
11.	*Terminalia - Anogeissus latifolia - Tectona series* (Deciduous Type)	8047.22	Gir WS & NP Junagarh Distt.	Gujarat
			Mudumalai WS, Nilgiri Distt.	Tamil Nadu
			Ganjan WS, Panna D istt.	MP
			Madeshwara Malai, Mysore Distt.	Karnataka
			Bandipur Tiger Resserve Mysore	-do-
			Distt. Nagarhole NP, Mysore Distt.	-do-
			Nilgiri eastern-slope Coimbatore Distt.	Tamil Nadu
			Talamalai, Coimbatore D istt.	-do-
12.	*Tectona - Terminalia series* (Deciduous Type)	14629.53	Bastar Distt.	MP (including Chhatisgarh)
			Indravati N P, Bastar Distt.	MP (including Chhatisgarh)
			Tadobaa NP, Chandrapur Distt.	Maharashtra
13.	*Tectona - Terminalia - Adina - Anogeissus series* (Deciduous Type)	602.70	Dang Distt.	Gujarat
14.	*Tectona - Dillenia - Lagerstroemia lanceolata - Terminalia paniculata series* (Deciduous Type)	7252.80	Anashi Ghat (Part of Dandeli WS) N Kanara Distt.	Karnataka

1.	2.	3.	4.	5.
16.	*Shorea - Cleistanthus - Croton series* (Deciduous Type)	393.01	Dalma WS Singhbhum Distt.	Bihar (including Jharkhand)
17.	*Shorea - Terminalia - Adina series* (Deciduous Type)	12944.09	Manas WS, Barpeta Distt.	Assam
			Palamau Tiger Reserve Daltanganj	Bihar
			Distt. Champaran	Bihar
			Kanha NP,	MP
			Mandla Distt.	-do-
			Corbett N P	Uttaranchal
			Dudhwa NP, Lakhimpur Distt.	Uttaranchal
			Neora Valley, Darjeeling Distt.	West Bengal
18.	*Shorea - Dillenia - Pterospermum series* (Deciduous Type)	-	-	-
19.	*Shora - Syzygium operculatum - Toona - Symplocos series* (Deciduous Type)	657.05	Simplipal WS & NP, Mayurbhanj Distt.	Orissa
20.	*Toona - Garuga series* (Semi evergreen Type)	-	-	-
21.	*Bridelia - Ficus glomerata - Syzygium series* (Semi evergreen Type)	112.60	Mt. Abu WS, Sirohi Distt.	Rajasthan
22 & 23.	Shola (Montane) Forest & Gordonia *Schefflera - Meliosma series* (Semi evergreen Type)	70.00	Ernakulam NP, Idukki Distt.	Kerala
			Anamalai & Tunacadu WS, Coimbatore Dist.	TN (Tamil Nadu)
			Ebanadu, Nilgiri Dist.	TN
			Kaubaatti, Nilgiri Distt.	TN
24.	*Memecylon - Actinodaphne - Syzygium series* (Evergreen Type)		Kankumbi, Belgaum Dist.	Karnataka
			Maharashtra Koyna valley, Satara Distt.	

1.	2.	3.	4.	5.
5.	*Persea - Holigarna - Diospyros series* (Evergreen Type)	-	-	-
6.	*Dipterocarpus - Mesua - Palaquium series* (Evergreen Type)	2066.40	Govardhangiri Attigod Jog Forests, Shimoga Dist.	Karnataka
7.	*Cullenia - Mesua - Palaquium series* (Evergreen Type)	2337.59	Agastyamalai kala Kadu WS, Kanyakumari Dist. Ervikulam NP, Idukki Dist. Thekkady Tiger Res. Silent Valley NP, Palghat Dist.	TN Kerala -do- -do-
8.	*Acacia-Albizia amara* (Dry E vergreen Type)	-	-	-
9.	*Manilkara - Chloroxylon series*	520.30	Point Calimere WS Thanjavur Dist. Pulicat Bird Sanctuary (wetland) Nellore Dist.	TN Andhra Pradesh
0.	*Albizia amara - Chloroxylon - Anogeissus latifolia series* (Dry Evergreen Type)	43.80	-	-
1.	Tropical wet evergreen forest (Himalayas)	1950.50	Danseri - kaki - Disama RF	Nagaland
2.	Tropical moist deciduous forest	1078.53	Intangki WS, Kohima Dist. Kaziranga NP, Jorhat Dist. Keibul Lamjao (Westland) NP, Central Dist. Balphakram NP	Nagaland Assam Manipur Meghalaya
3.	Sub-tropical broad leaved hill forest	-	-	-

1.	2.	3.	4.	5.
34.	Sub-tropical evergreen sclerophyllous forest	110.30	-	-
35.	Sub-tropical *Pinus roxburghii* forests (Himalaya)	1092.09	-	-
36.	Temperate mixed oak & coniferous forest (Himalayas)	1555.82	Kistwar high altitude NP, Doda Dist.	J.&K. (Jammu & Kashmir)
37.	Temperate coniferous forest (Himalayas)	656.96	Dachigam NP, Srinagar Dist.	J.&K.
			Gulmarg M&B Reserve, Srinagar Dist.	-do-
			Overu-Aru M&B Reserve	-do-
38.	Wet temperate forest (Himalaya)	410.49	Namdapha WS, Tirap Dist.	Arunachal Pradesh
39.	Sub-alpine altimontane forest (Himalaya)	1101.48	Sainj-Tirtham valley,	HP
			Kedarnath WS, Chamoli Dist.	Uttaranchal
			Nandadevi NP	Uttaranchal
			Henus high altitude NP Leh Dist.	J.&K.
40.	Alpine scrub & meadows (Himalaya)	408.83	Valley of Flowers, Chamoli Dist.	Uttaranchal
41.	Alpine steppe (Himalaya) Leh Dist.	1195.09	Shang Gaui WS Reserve	J.&K.
			Henus high altitude NP	J.&.K.
			Overa-Aran M&B Res., Srinagar Dist.	J.&.K.
			Kistwar high altitude NP, Doda Dist.	J.&K.
			Sinalila NP, Darjeeling Dist.	W.B.(West Bengal)

1.	2.	3.	4.	5.
42.	Mangrove	8158.39	Sunderban Tiger Reserve, 24 Parganas Distt. (West Bengal)	W.B.
			Point Calimere WS,	TN
			Wild Ass Sanctuary, Surendranagar Distt.,	Gujarat
			Killai, Cuddalur Distt.	TN
			Pachavaram,Cuddalur	TN
			Andaman-Nicobar Is.	A&N. (Andaman & Nicobar)
43.	Tropical wet evergreen forest (Andaman & Nicobar) (main Island)	811.18	Saddle peak NP, N Andaman Dist.	A&N
			Jarawa area NP, S & Middle Andaman (East)	-do-
			North Button NP (Middle Andaman)	-do-
			South Button NP	-do-
			Mount Hariet NP, South Andaman	-do-
			Interview Is.	-do-
			Tar Mugli group of Coral Reef Islands	-do-

Gadgil and Meher - Homji (1986) defined as many as 43 different vegetation types ranging from *Calligonum* series + *Prosopis - Salvadora Capparis Ziziphus* series of Jaisalmer Dist. in Rajasthan to *Cullenia - Mesua - Palaquium* series in Silent Valley National Park in Palaghat District of Kerala, and from Temperate coniferous forest type of Dachigam National Park of Jammu & Kashmir Himalaya to Tropical wet evergreen forest type of Andaman & Nicobar Islands and Tropical moist deciduous forest type of Balaphakram, Keibul Lambjao (Wetland) and Kaziranga National Parks in Meghalaya, Manipur and Assam respectively.

Ten biogeographic regions (including 3 basic biomes and 2 natural realms) cover Himalayan Highlands, Thar Desert, Malabar Rain Forest, Indo-Ganga Monsoon Forest, Deccan Thorn Forest, Coromandel, Mahanadian, Bengalian Rain Forest, Andaman & Nicobar and Lakshadweep Islands. As per vegetation types and plant diversity 12 different biogeographic zones

Biogeographic specific Hotspots of Biodiversity, Threatened and Spectacular Plants of Indian Mainland

Sl.No.	Hot spots	Biogeographic zone
1.	Karakoram & Ladakh	Trans Himalaya
2.	Kumaon-Garhwal Himalaya	W. Himalaya
3.	Siwaliks	W. Himalaya
4.	Sikkim Himalaya	E. Himalaya
5.	Arunachal Pradesh	E. Himalaya
6.	Lushai hills	E. Himalaya
7.	Tura, Balphakram, Khasi Hills (Meghalaya)	N.E. India
8.	Aravallis	Semiarid zone
9.	Bundelkhand	Gangetic plain
10.	Chotanagpur plateau	Deccan
11.	Panchmarhi-Satpura ranges	Deccan
12.	Simlipal & Jeypore hills Orissa	Deccan
13.	Bastar & Koraput hills	Deccan
14.	Vishakhapatnam hills & Araku Valley	Deccan
15.	Tirupati-Cuddappa hills	E. Ghats
16.	Marathwada hills	Deccan
17.	Saurashtra-Kutch	Deccan
18.	Mahabaleshwar-Khandala ranges	W. Ghats
19.	Agumbe-Phonda ranges	W. Ghats
20.	Ratnagiri & Kolaba ranges	W. Ghats
21.	Nilgiris	W. Ghats
22.	Silent Valley & Wynaad	W. Ghats
23.	Anamalai	W. Ghats
24.	Idduki-Sabarigiri	W. Ghats
25.	Kalakad & Agastaimalai hills	W. Ghats

Status of Hotspots

S.N.	Biogeographic Zone	No. of Hotspots
1.	Trans Himalaya	1
2.	W. Himalaya	2
3.	E. Himalaya	3
4.	NE India	1
5.	Gangetic Plains	1
6.	Deccan	7
7.	E. Ghats	1
8.	W. Ghats	8
9.	Semi arid zone	1
	Total	25

Giant Trees of India (Bhadran 1958)

Species	*Girth*	*Height*
Abies densa	274.32 cm	30.48 m
A. pindrow	498 cm	62.8 m
Acacia nilotica	300 cm	25 m
Acacia senegal	80 cm	6 m
Acer campbelli	392.4 cm	27.4 m
Acer oblongum	26 cm	34 m
Adina cordifolia	782 cm	41.8 m
Aesculus indica	6 m	44 m
Ailanthus excelsa	284.48 cm	25.29 m
Albizia procera	419.1 cm	42.36 m
Alstonia scholaris	300 cm	20 m
Betula alnoides	274.32 cm	25.90 m
Bombax ceiba	500 cm	45 m
Callophyllum tomentosa	388.62 cm	34 m
Cedrus deodara	1646 cm	73.2 m
Chukrassia velutina	281.94 cm	23.16 m
Cupressus torulosa	881.38 cm	56.36 m
Dalbergia sissoo	250 cm	40 m
Diospyros melanoxylon	269.24 cm	-
Dipterocarpus gracilis	853 cm	47.2 m
D. griffithii	548.64 cm	43.28 m
D. incanus	670.56 cm	-
D. macrocarpus	140 cm	60 m
D. turbinatus	723.9 cm	47.24 m
Emblica officinalis	350 cm	20 m
Exbucklandia populnea	571.5 cm	30.48 m
Fraxinus floribundus	150 cm	15 m
Gmelina arborea	450 cm	30 m
Hardwickia binata	300 m	22 m
Holoptelea integrifolia	594 cm	53 m
Juglans regia	700 cm	42 m
Madhuca indica	350 cm	25 m
Mangifera indica	685.8 cm	36.5 m
Mesua nagasserium (ferrea)	320.04 cm	-
Michelia champaca	350 cm	40 m
Mimusops elengi	300 cm	18 m
Mitragyna parviflora	300 cm	15 m
Picea smithiana	752 cm	65.5 m
Pinus roxburghii	405 cm	64.5 m
P. wallichiana	335.28 cm	41.14 m
Platanus orientalis	1249.68 cm	-
Pterocarpus dalbergioides	601.98 cm	46.02 m
P. marsupium	330.2 cm	35.35 m
Quercus himalayana	342.90 cm	41.14 m

Species	*Girth*	*Height*
Schleichera olesoa	373.38 cm	32.30 m
Shorea robusta	450 cm	38 m
Tamarindus indica	600 cm	25 m
Tetrameles nudiflora	1615.44 cm	36.57 m
Terminalia arjuna	1097 cm	30.5 m
T. bellirica	400 cm	38 m
T. chebula	240 cm	25 m
T. myriocarpa	600 cm	41 m
Toona ciliata	300 cm	30 m
Tsuga brunoniana	619.76 cm	36.57 m
Ulmus wallichiana	1292.86 cm	-
Ziziphus mauritiana	90 cm	12 m

Source: Bhadran (1958)

representing Trans - Himalaya, West Himalaya, East Himalaya, NE India, The Indian Desert, Semi-arid zone, Gangetic plain, Western Ghats, Deccan Peninsula, Indian Coasts, Andaman & Nicobar Islands and Lakshadweep Islands of phytogeographical significance are classified. These zones provide safe haven to the rare and threatened elements. Based upon the occurrence of endemic elements and species richness 25 major 'hot spots' are been identified ranging from Karakoram & Ladakh in Trans Himalayan zone and Kumaon -Garhwal Himalaya in western Himalaya to Aravallis in Semiarid zone to Agumbe-Phonda Ranges in western Ghats covering as many as nine biogeographic zones.

Biswas *et al.* (1996) while working on the threatened medicinal plants of Indian Himalaya proposed following sites for habitat preservation, *in situ* conservation and range extensions. These areas are suitable for the conservation of general biological diversity also.

Jammu & Kashmir: Udhampur & Pathankot Ranges, Higher reaches of Budharboo, Doskit, Nubra valley (Ladakh), Srinagar valley, Upper ridges of Kashmir valley etc.

Himachal Pradesh: Koti-Gulaba Range, Parbati valley, Dharamsala Division (Banoi Range, Macleodganj environ), Dhauladhar Range etc.

Uttaranchal: Barhat Range (Uttar Kashi), Manora (Nainital), Mamuna Shivpur Range (Mussoorie hills), Chakrata, Ukhimath, Bhilangana Range, Kotikanasar (Chakrata Division), Deoban Ridge & shaddier aspect of Surkunda Devi Hills (Tehri Division), Badkot, Dhanaulti & Gangnani Ranges, Mandal Forests, Palam valley etc.

North Bengal Himalaya (West Bengal) Kurseong range, Sanglila, Neera, Jorepokhri, Senchal, Buxa ranges, Sukna range (Rajabhatkhowa), Kalimpong, Tiger Hill and Jalpaiguri ranges, Duars etc.

Floristic richness point of view western Ghats, northeastern region alongwith eastern Himalayas, and Andaman and Nicobar Islands of India form two of 25 major hotspots of the globe. Over 40% of flora and 30% of vestebrate fauna are confined to these hotspots (Biswas, 2000).

Sikkim: Tista valley, Sevoke, Lachen valley, Fambong, Maenam, Singba ranges, Rungit valley, Kanchenjunga, Indo-Bhutan border, W & E Sikkim ranges, Indo-Nepal and Tibet borders etc.

Arunachal Pradesh: Dibang valley (Subansiri) Mahao Range, Tenga valley, Banderdeva Division, Bomdila range, Kameng Dist., Patkoi range, Pakhui, Walong and Itanagar ranges etc.

31 microhabitats representing priority centres for biodiversity and forest genetic resources in different parts of Indian region are well identified (Gupta and Biswas, 1997). These microhabitats cover Ladakh, Kedarnath, Nanda Devi, Siwaliks, Terai in Northern India to North, South and Nicobar Islands, and from Saurashtra Cutch in Gujarat to Namdapha and Lushai Hills in North eastern India. Studies carried out on these micro-habitats reveal useful data on not only the species richness but also different suggestive measures for *in situ* conservation.

Preservation of Biodiversity in Ecotones

Removal of ecotones between the two or more land use systems has resulted into the decline of biodiversity and interdependent elements of significance of gene flow. Ecotones of significance of different sizes and vegetation types occur in different parts of the country. The preservation of these sites can help in conservation of species.

Ex situ Conservation of threatened Forest Flora

Botanic Gardens:

Worldwide there are over 1500 botanic gardens engaged in plant conservation works. India has 12 Botanic Gardens, out of which about five are represented by threatened species. Botanic Garden of Forest Research Institute has numerous threatened elements of national and international importance. Among the rare and endangered species of forest genetic resource of international importance, the species are *Tectona hamiltoniana* Wall., *Araucaria cunninghamii* Aiton ex Lambert, *Eucalyptus deglupta*, *Juniperus bermudiana* Linn., *J. procera* Hochst ex Endl. *Pinus pseudostrobus* Lindl., *Platanus orientalis* Linn., *Khaya senegalensis* (Derr.) A.Juss. (see FAO Forestry Paper No.77, 1986).

Lloyd Botanic Garden, Darjeeling (West Bengal) has threatened taxa of different conifers, Magnolias, maples, oaks etc. representative of eastern Himalaya. Indian Botanic Garden, Howrah under Botanical Survey of India has genetic resource of palms, bamboos, aquatics, jasmine, cacti, orchids etc.

National Botanical Research Institute, Lucknow (NBRI), Uttar Pradesh with its adjunct gardens in its research and development pursuit carries out the activities such as introduction, conservation, propagation, genetic upgrading and utilization of indigenous and exotic species. The institute carries out studies on the conservation of threatened species of northern India such as *Sophora mollis, Eremostachys superba, Phoenix acaulis* etc. NBRI has initiated works on the taxonomic-cum-genetic characterization of *Aisandra butyracea* (Cheura), an edible oil yielding forest species of Uttaranchal.

Jaipur in Rajasthan is reported to have representation of species of Wattles (*Acacia*) from different parts of the world in a recently established World Arboretum near Jaipur in Rajasthan (Abhijit Ghosh *pers. communication*).

Government Botanic Gardens, Ootacamund (Tamil Nadu) is represented by rich diversity of tree genetic resource of typical of hill tops of South India. Among the important species are *Cedrus deodara, Rhododendron nilagirica, Quercus montana, Pinus wallichiana, Magnolia, Saurauja, Aesculus, Alnus* spp. *Albizia, Acacia, Yucca, Mahonia, Eugenia* etc.

It deserves a mention here that many plant species of economic value believed or presumed to be extinct in wild and/or from the face of the globe have been found growing in some of the botanical gardens of the world. Thus the gardens play very significant role in conserving the genetic resource. There is need for inventorization of relict gardens of great history of plant introduction.

Gene Conservation

Cryopreservation and *in vitro* ***storage***

This includes the maintenance of cells, tissues or organs in cultures where growth is slowed technically. The cryopreservation and *in vitro* storage may have some application as a backup conservation strategy. Cryopreservation may be quite useful in threatened conifers where rejuvenation is difficult (Biswas 1999). Preservation of reproductive germplasm through the encapsulated artificial seed i.e., embryonic material is becoming an important viable option.

Molecular markers and genetic fingerprinting:

For diversity evaluation at infrageneric and intraspecific levels variations in cellular molecules such as DNA and proteins are very useful compared to environmentally affected features such as quality, vigour and different morphological features (Biswas 1995a). Molecular markers can prove quite significant in establishing the taxonomic entity and trekking the genetic relationship among the threatened taxa. A team of international scientists in United Kingdom launched a project (Bar code of Life Initiative) to identify genetically, or provide a bar code for every plant and animal species. This would be based upon the short DNA sequence to be taken from the organism and that can characterize the form. This would also work for rapid identification of species. The Initiative has projects aiming at 8000 species of plants typical of Cost Rica and Central America, 10,000 species of known birds and 23, 000 kinds of marine and fresh water pisces (*Times of India* Daily, 10.February 2005).

Field Gene Banks

Field Gene Bank is a representative site with widest practicable range of genetic diversity at infraspecific or infrageneric level assembled for conservation purposes. For the conservation evaluation of threatened flora a field gene bank plays a very significant role. Many of the Forest Research Institutes world over maintain diversity of different forest tree genetic resource. Forest Research Institute, Malaysia maintains over 700 woody species particularly the commercially important dipterocarps from Southeast Asia.

Seed Banks

Seeds of different species carry with them diversity of genes. Maintenance of a seed bank involves uninterrupted supply of power, monitoring, testing of viability and introduction within a stipulated time. According to Groombridge (1992) nearly 50,000 plant species produce seeds that do not survive low temperature and/or dehydration. Recalcitrant seeds of many tropical species possess no natural dormancy and die soon if not allowed to germinate quickly. Such seeds and the plants are required to be maintained *ex situ*. Botanic gardens with a low temperature seed storage facility are very few in India. The country has 2 seed storage facility compared to 9 in Japan, 11 in France, 10 each in Germany and United Kingdom, 28 in USA, 6 in Australia (Heywood *et al.* 1990). Many of these vary from regional to international in scope.

Reintroduction

Reintroduction or rehabilitation of species is a high risk conservation strategy. Although the plants are multiplied but their reintroduction to the areas of their original habitat or similar congenial ecological niche has not been given due emphasis. Suitability of the habitat or bioclimatic zone for reintroduction of species needs support of inventory and integration of remotely sensed data and GIS (Geographical Information System) database (Biswas, 1999).

Noble Hardwoods or Giant Trees Preserved in India

The country has several species of trees preserved in different parts of forest and modified ecosystem. The individual stands have characteristics of dendrochronological significance.

Establishment/strengthening of regional to national level networking on monumental Preservation Plots or Giant Trees Network on the lines of

Noble Hardwoods Network of European Forest Genetic Resources Programme (EUFORGEN) is required to be made in order to conserve the genetic variability and phenotypic variation among the noble hard woods of Indian forest flora.

Literature Cited

Ali, S.I. and Quiser, M. (1980): Hybridization in *Acacia nilotica* (*Mimosoideae*). *Bot. J. Linn. Soc.* 80 (1):69-77.

Anon. (1989): *Annual Research Report* - 1989. Forest Research Institute, Dehra Dun.

Anon. (1993): *Annual Report*, 1992-93. Indian Council of Forestry Research and Education, Dehra Dun.

Anon. (1994): Annual Report, 1993-94. Indian Council of Forestry Research and Education, Dehra Dun.

Arora, R.K. and Engels, J.M.M. (1993). Genetic resources in medicinal and aromatic plants and their conservation and use. *Acta Horticulturae* 330:21-38.

Bahadur, K.N. (1988): Monograph on genus *Toona*. Bishen Singh & Mahendra Pal Singh, Dehra Dun.

Banik, R.L. (1995): Diversities, reproductive biology and strategies for germplasm conservation of bamboos. In: VR Rao & AN Rao (Eds.) "*Bamboo and Rattan Genetic Resources and Use*." *Proceed. Ist INBAR Biodiversity, genetic resources and conservation working group*, 7-9 November 1994, Singapore: 1-4.

Bhadran, C.A.R. (1958): Giant trees of India. *Indian Foreter* 84 (6):318-333.

Biswas, Sas. (1990): Ethnobotanical studies on northeastern region of India with particular emphasis on conservation of biological diversity. *Proceed. 2nd International Congress of Ethnobiology, "The Challenges of Ethnobiology in the 21st Century*" held at Kunming, China, 22-26 Oct. Symp. Session:128.

Biswas, Sas. (1991): Rare and endangered flora of eastern Himalaya and the measures for its conservation. In: GS Rajwar (Ed.) *Advances in Himalaya Ecology* 6:267-273. Yesterday, Today & Tomorrow Pub., Delhi.

Biswas, Sas. and Dayal, R. (1994): Endangered biodiversity of Indian flora and strategies for its conservation. In: K.C. Sahni (Ed.) Higher plants of Indian sub-continent III: 281-287. IBD. Pub., Dehra Dun.

Biswas, Sas. (1995): Diversity and genetic resource of Indian bamboos and the strategies for their conservation. In: VR Rao & AN Rao (Eds.) "*Bamboo and Rattan Genetic Resources and Use*". *Proceed. Ist INBAR biodiversity, genetic resources and conservation working group*, 7-9 November, 1994, Singapore: 1-4.

Biswas, Sas. (1995a): Genetic management plans and genetic diversity measurement. *In*: Final Report on Tree Improvement. Fellowship awarded by FAO of United Nations for training at the Institute of Forest Genetics, Grosshansdorf, Germany: 2-46.

Biswas, Sas.; Chandra, S. and Jain, S.S. (1996): Rare and Endangered bamboos, rattans, and medicinal plants of Indian Himalaya and the strategies for their conservation. *Proc. International workshop* on "*Role of bamboos rattans and medicinal plants in mountain development*", 15-18 May, 1996: 1-29. organised by INBAR, IPGRI, IMPN and ICMOD Pokhra, Nepal.

Biswas, Sas. and Wuehlisch, Georg von (1997): Identification of *Populus nigra* Linn. clones by their isozymes with particular reference to Germay. *Proceed. 84th Session of Indian Science Congress*, Delhi 1997, *Section of Botany*. Adv. notes for Symp. on Biotechnology: 21.

Biswas, Sas.; Pathak,KC; Bhuyan, TC; Ahmed, A. and Thakuriya,GC (1997): Flora of Burnihat Valley, Assam - Meghalaya. *Addl. Series of Indian J. Forestry*, Dehra Dun.

Biswas, Sas. (1999): Conservation status of threatened Indian potential forest genetic resource and need for diversity and distribution evaluation. *In*: Negi, RS *et al.* (Eds.) "*Forestry Research in Conservation of Natural Forests*": Proceedings of an Intl. Workshop UNDP - ICFRE Project IND/92/038 India: 33-41. ICFRE.

Biswas, Sas. (2000). Threatened hotspots of Indian forest biodiversity - need for conservation of floristics. *Van Vigyan* 38(1-4):90-104.

Champion, H.G. & Seth, S.K. (1968): Revised survey of forest types of India. Manager of Publication, Govt. of India, New Delhi.

Chatterjee, D. (1940): Studies on the endemic flora of India and Burma. *J. Asiat. Soc. Bengal* (*n.s.*) 5:19-67.

Devy, MS and Ganesh, T (2003). Canopy science and its relevance in India. *Current Science* 85(5):581-584.

Dogra, P.D. (1981): Forest genetics - Research and Application in Indian Forestry-I. *Indian Forester* 107(4):191-219.

Dogra, P.D. (1986): Species diversity and gene conservation in Indian tree species in relation to forestry. *Indian Forester* 112(7):598.

Food and Agriculture Organizations of the United Nations (FAO) (1979): Databook on endangered tree and shrub species and provenance. FAO Forestry Paper 77: pp. 524. Rome.

Forest Survey of India (FSI) (1996): The state of Forest Report - 1995 pp. 98. FSI, Dehra Dun.

Forest Survey of India (FSI) (2000): State of Forest Report 1999. pp. 113. FSI, Dehra Dun.

Gadgil, M and Meher-Homji, VM (1986): Localities of great significance to conservation of India's biological diversity. *Proc. Indian Acad. Sci.* (Animal/ Plant Sci.) Suppl.: 165-180.

Groombridge, B. (1992): Global biodiversity-status of earth's living resources. A report compiled WC Honitorey Centre, Chapman & Hall, London.

Gupta, B.N. and Totey, N.G. (1994): An account of preservation plots in Central India. *Indian Forester* 120 (9):844-859.

Gupta, B.N. (1997): Biodiversity of Indian Forests - some analytical facts. Proceed. Nat. Seminar on Biological Diversity & Human Welfare Govt. PG College Rishikesh 28-30 Sept. 1997: 1-8.

Gupta, B.N. and Biswas, Sas. (1997): Biodiversity characterisation at landscape level using satellite remote sensing. *Proceed. "Biodiversity characterisation using satellite remote sensing*" Project of National Remote Sensing Agency, Hydrabad.Dept. of Space & Dept. of Biotechnology, Govt. of India.

Gupta, G.C.; Sharma, K.C. and Singh,C. (1970): Programme for breeding of high resin yielders of chir pine in Himachal Pradesh. *Proceed. Seminar/Workshop.* FRI & Colleges Dehra Dun: 46-50.

Heywood, C.A.; Heywood, V.H. and Wyse Jackson, P. (1990): International Directory of Botanical Gardens V, 5th ed. Koeltz Sc. Books.

Jain, S.K. (1990): Botanical regions, and flora of India. *Everyman's Science* 24 (6):213-221.

Kemp, R.H. (1992): The conservation of genetic resources in managed tropical forests. *Unasylva* 43 (169):34-40.

Khullar, P. (1992): Conservation of biodiversity in natural forests through Preservation Plots - a historical perspective. *Indian Forester* 118 (5):327-337.

Myers, Norman (1981): Conservation needs and opportunities in tropical moist forests. *In*: H. Synge (Ed.). The Biological aspects of rare plant conservation. John Wiley & Sons Ltd.

Nair, P. Vijayakumaran and Menon, ARR (1998): Estimation of bamboo resources in Kerala by remote sensing techniques. *Current science.* 75 (3): 209.

Namkoong, G. (1986): Genetics and the future of the forests. *Unasylva* 38 (152):2-18.

Pearson, R.S. and Brown, H.P. (1932): Commercial timbers of India. Central Pub. Branch, Calcutta.

Prance, G.T. (1984): Completing the inventory. *In*: Heywood, V.H. & Moore DM (Eds.)): *Current Concepts in Plant Taxonomy*: 365-396.

Prasad, S. Narendra; Vijayan; Balachandran, S; Ramachandran, VS and Verghee, CPA (1998): Conservation planning for the western Ghats of Kerala: I. A GIS approach for location of biodiversity hot spots. *Current Science* 75 (3): 211.

Rodgers, W.A. (1991): Forest Preservation plots in India- Management Status and Value. *Indian Forester* 117(6):425-433.

Roy, P.S. and Sarman Singh (1995): Biodiversity conservation - Role of remote sensing and GIS. *Tech. Report.*: 1-22. Indian Remote Sensing Institute, Dehra Dun.

Roy, P.S.; Tomar, S. and Jaganathan, C. (1997): Biodiversity characterisation at landscape level using satellite remote sensing. *NNRMS Bull.* (B)-21: 12-18.

Sanchez-Azofeifa, A. (2002) *In*: Mitchell, A. *et al.* (Eds.) Global Canopy Handbook; 196-199, Global Canopy Foundation, Oxford, UK.

Sastry, A.R.K. and Nayar, M.P. (Eds.) (1987-90): Red Data Book on Indian Plants. Vol.I-III. Botanical Survey of India, Howrah.

Swaminathan, M.S. (1993): Agriculture cannot wait! *Everyman's Science* 28 (1):17-24.

Tewari, D.N. (1992): A monograph on teak (*Tectona grandis* Linn.f.). International Book Distributors, Dehra Dun.

Tewari, D.N. (1992a): Biodiversity and forest genetic resources. International Book Distributors, Dehra Dun.

Tewari, D.N. (1995): Forests, Gardens, Parks and Urban Environment. pp. 370. IBD. Dehra Dun.

Troup, R.S. (1921): The silviculture of Indian trees. Clarendon Press, London.

Vavilov, N.I. (1951): The origin, variation, immunity and Breeding of cultivated plants. *Chronica Botanica* 13:1-364.

Wuehlisch, G. von; Chauhaan, P.; Biswas, Sas.; Ahuja, M.R. and Muhs, H.J. (1996): Vermehring and charakterisierung von S chwarz pappel - Altabaumen zur Generhaltung. (Multiplication and characterisation of old trees of black poplar for gene conservation) *In*: (Begemann, F., Ehling, C. & Falge, R.) Vergleichende Aspekte der Nutzung und Erhaltung pflanzen-und tiergene-tischer Resourcen. Band 5 (Nutztiere) Tagungs band eines Symposiums vom, 7-9 Oct. 1996 in Mariensee: 344-346.

PLATE III

Threatened Plant Diversity of India

1. *Phaius tankervillae*- ground orchid
2. *Paphiopedilum fairrienum* once a 'Lost Orchid'
3. *Cycas cercinalis* with male flowers
4. Flowering of *Eremostachya superba*

Criteria and Indicators for Threatened Biodiversity Categorization

The different rare and threatened categories being followed today have been used for the past over 30 years. Biologists and conservationists use the words such as "*extinct*', "*endangered*", "*vulnerable*", "*critically endangered*" and "*extinct in the wild*" across the world. To focus attention on the strategies for conservation the categories highlight those species under higher extinction risk. The categories apply well to the ecogeographic, phyto/zoogeographic wild populations inside their natural ranges and infraspecific taxa.

Threatened species categorization

On an international level, the most widely used criteria are the 'Red Data Book Categories' of IUCN having a relatively subjective measure of the degree of threat to individual species. These are more widely applicable than the more fixed criteria (Synge, 1981). The categories typically apply to species and can apply to infraspecific taxa as well (Moulton and Sanderson, 1997). As per the threats operative on the taxa the species are scheduled and applicable to the act of Wildlife Preservation.

The categories and definitions according to the Red Data List (IUCN 1994) are as follows:

I. **Extinct (Ex)**: A taxon is extinct when there is no reasonable doubt that the last individual has died.

II. **Extinct in the Wild (EW)**: A taxon is extinct in the wild when it is known only to survive in cultivation or domestication or as a naturalized population (or populations) well outside the past range of historical distribution.

III. **Critically Endangered (CR)**: A taxon is critically endangered when it is facing an extremely high risk of extinction in the wild in the immediate future.

IV. Endangered (EN): A taxon is endangered when it is not critically endangered but is facing a very high risk of extinction in the wild in the near future.

V. Vulnerable (VU): A taxon is vulnerable when it is not critically endangered or endangered but is facing a high risk of extinction in the wild in the medium-term future.

VI. Lower Risk (LR): A taxon is lower risk when it has been evaluated and does not satisfy the criteria for any of the categories critically endangered, endangered, or vulnerable. Taxa included in the lower risk category can be separated into three sub-categories:

1. *Conservation Dependent* (CDS) - Taxa that are the focus of a continuing species - specific or habitat-specific conservation program targeted toward the taxon in question, the cessation of which would result in the taxon qualifying for one of the threatened categories mentioned within a period of five years.

2. *Near Threatened* (NT) - Taxa which do not qualify for conservation dependent, but which are close to qualifying for vulnerable.

3. *Least Concern* (LC) - Taxa which do not qualify for conservation dependent or near threatened.

VII. Data Deficient (DD) - A taxon is data deficient when there is inadequate information to make a direct or indirect assessment of its risk of extinction based on its distribution and/or population status. A taxon in this category may be well studied and its biology well known, but appropriate data on abundance and/or distribution are lacking. Data deficient is, therefore not a category of endangered or lower risk. Listing of a taxon in this category indicates that more information is required and acknowledges the possibility that future research will show that threatened classification is appropriate It is important to make positive use of whatever data are available. In many cases, great care should be exercised in choosing between data deficient and threatened status. If the range of a taxon is suspected to be relatively circumscribed and a considerable period of time has elapsed since the last record of the taxon, threatened status may well be justified.

VIII. Not Evaluated (NE) - A taxon is considered not evaluated when it has not yet been assessed against any of the categorizing criteria.

IX. Threatened (T): Used in the conservation context for species which are in one of the three categories, namely Critically endangered, Endangered and Vulnerable. Some species are marked as threatened where it is known that they are Endangered, Vulnerable or Rare, but there is not enough information to say which of the three categories is appropriate.

X. Rare (R): Taxa with small world populations that are not at present endangered or vulnerable, but are at risk. These elements are usually localized within restricted geographical areas or habitats or are thinly scattered over a more extensive range.

XI. Indeterminate (I): Insufficient information is presently available as regards to any of the categories (Rare, Endangered, Vulnerable) the taxon falls.

XII. Reproduction Constraints (REP): Taxon with diagnostic features on the reproductive parts influencing the survival falls under REP.

XIII. Out of danger (O): Taxa formerly included in one of the categories, but which are now considered relatively secure because effective conservation measures have been taken or the previous threat to its survival has been removed.

The norms of categorization are likely to differ while considering the threat status on the basis of site-cum-species specific aspects. In a situation where the area is small and the species is governed by its genetic and reproductive nature the count of the individual per unit area may not suffice.

Threatened Habitat Categorization

Although criteria and indicators for different categories of threat with respect to species are well-outlined and defined but there is need for categorization of habitat threatened due to varying impacts. Biswas (2003) believes that ecological-cum-in floristic niche of biodiversity importance becomes endangered due to varying degree of threats. It is found that the status of habitat with respect to different locality factors including physiographical, climatological and edaphic contribute towards not only to the development of species but restoration or regeneration also. Biswas (2003) proposed categories of threatened habitats of biodiversity significance for international acceptance on the basis of development of models, criteria and indicators for conservation of indigenous flora (relict and disjunct types) of biodiversity of a site/region, restoration of fragmented ecosystem of phytogeographic significance, bioaesthetic-cum-biodiversity backed eco-restoration, creation of artificial biosphere in threatened sites, restoration of rare and threatened preservation plots, integration of indigenous traditional knowledge on the threatened habitats for eco-restoration and sustainable management, and removal of invasive/weeds of threatened habitats. The categories considered are as following:

1. Extinct (E): Habitat is extinct when there is no scientific evidence about its existence. For example, biodiversity rich sites along the coasts and islands engulfed as a consequence of tsunami waves in Andaman and Nicobar Islands of India after 26th December 2004 are considered extinct.

2. Critically endangered habitat (CRH): A habitat is critically endangered when it faces ever increasing impacts which ultimately results into the extinction of the habitat in immediate future. As observed there is reduction shrinkage of at least 80% of the habitat over the last 20 years or suspected further on account of the impact of vulnerable environmental or geomorphological stress. This also shows no chance of restoring or extending the habitat. Examples of CRH are shrinking wetland or swampy sites and site invaded by weeds and eroding at the sametime. The probability of extinction is at least 50% within 20 years.

3. Endangered habitat (EH): Not critically threatened but facing a high risk of extinction in the near future. This is based on the bioloical and environmental aspects of threats operative on the habitat. There is suspected reduction or shrinkage of at least 50% over the last 10 years. As subjected to biodiversity there is need to consider the status of biological diversity and interdependent elements of the habitat. Quantitative analysis which is to include the habitat viability should show extinction of at least 20% within 30 years.

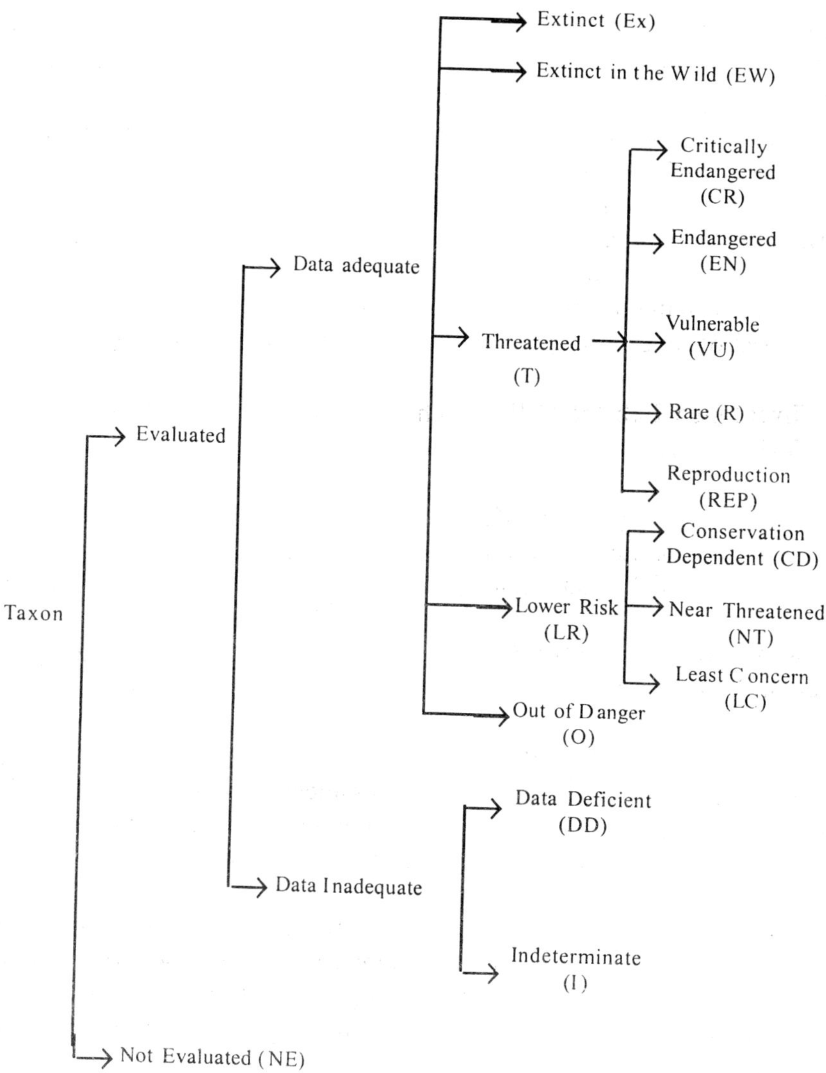

STRUCTURE OF THE CATEGORIES

4. Vulnerable habitat (VH): Habitat facing risk of extinction in the medium term future. There is decline or shrinkage of at least 20% over the last 10 years under potential impacts both natural and man-made as well with extreme fluctuation on the biodiversity of as many as 10 locations typical of the region and the environment. The probability of extinction within less than 70 years subjected to the impacts.

5. Rare (RH): Such habitats are usually localized within restricted bioclimatic or biogeographic zone developing microhabitat and biological diversity which are not at present endangered or vulnerable, but at risk. RH may be under the impact of severe environment which includes climate change.

6. Threatened habitat (TH): Such habitats are one of CRH, EH, RH and VH.

7. Restoration Dependent Habitat (RDH): Such category is the focus of continuing habitat specific reclamation and preservation program, the cessation of which would place the habitat under one of the threatened ones.

8. Least Concern Habitat (LCH): Such habitats do not qualify for reclamation or eco-restoration and preservation-oriented activities.

9. Data Deficient Habitat (DDH): Such habitats are worthy for environmental or ecological impact assessment studies. Such categories require more information through direct or indirect assessment of its risk of extinction based on the inter-dependent components of biodiversity.

10. Restoration or recharging Constrained Habitat (RCH): Such habitats based on biophysical and climatological and other diagnostic features have restoring/recharging and rejuvenating capability, thus influencing the existence. Such habitat when environmentally impacted may enter into one of the threatened categories.

11. Out of danger (OH): Habitat once under one of the threatened categories but it is presently considered devoid of any threats. In biodiversity context such category has undergone effective and restorative measures with indigenous and naturalized flora, fauna and interdependent elements.

Literature Cited

Biswas, S as. (2003): E co-restoration of threatened habitats o f biodiversity significance. In: Proceed. Intl. Conference on Ecorestoration, 14-21 Oct. 2003: 143, National Institute of Ecology, New Delhi.

IUCN (International Union for Conservation of Nature and Natural Resources) (1994): Red Data List. Glands, Switzerland.

Mace, GM and Lande, R. (1991): Assessing extinction threats towards a revaluation of IUCN threatened species categories. *Conservation Biology* 5(2): 48-157.

Mace, GM; Collar, N; Cooke, J. Gaston; K, Ginsberg, J.; Leader Williams, N; Maunder, M and Milner-Gulland, EJ (1992): The development of new criteria for listing species on the IUCN Red List. *Species* 19:16-22.

Moulton, MP & Sanderson, J. (1997): Wildlife issues in a changing world. St. Lucie Press, Florida, USA.

Synge, H. (Ed.) (1981): The biological aspects of rare plant conservation. John Wiley & Sons, New York.

Chapter VI

Assessment and Evaluation of Threatened Flora of some Countries - an Overview and Case Studies

Assessing which species are threatened is only the beginning. For practical conservation of rare and threatened biodiversity and gene resource following levels are important: (i) assessment and evaluation of threatened taxa (ii) locating sites through literature and herbarium surveys (iii) field surveys for site and species-specific conservation (iv) monitoring the populations, and (v) proposing measures of conservation.

Attention on threatened plants was focussed way back in 1974 when Survival Service Commission (SSC) of International Union for Conservation of Nature and Natural resources (IUCN) set up the Threatened Plants Committee (TPC) with a view to providing information on botanical data for the c onservation o f t hreatened e lements to the g overnments to incorporate in their national programme.

Studies on rare and threatened species began when the principles of nature consrvation were clearly set forth in 1947 in the Report of the Wild Life Conservation Special Committee, entitled, "*Conservation of Nature in England and Wales*" with emphasis on conservation and maintenance, survey and research, experiment, education and amenity (West, 1981).

Assessment and evaluation of the threatened flora of different parts of the world have been on the forefront of several countries, such as that of United States (Ayensu 1978, 1981 Ward, 1979, Siddall *et al.* 1979, Fay, 1981); New Zealand (Given 1981), Australia (Good and Lavarack 1981), erstwhile USSR (Beloussova and Denissova, 1981), Sweden (see Nilsson 1981), Federal Republic of Germany (Sukopp and Trautmann, 1981), United Kingdom (Crompton, 1981; Perring and Farrell, 1977), European countries under European Economic Community (EEC) (Leon, 1980; Lucas and Synge 1981), Hawaii (Fosberg and Herbst, 1975), Japan (Shimizu and Satomi, 1976) Southern Africa (Hall *et al.* 1980), India (Jain and Rao, 1983; Sastry and Nayar, 1987).

United States of America

Assessment criteria of the US Federal Government is based upon following factors namely, 1. the present or threatened destruction, modification, or curtailment of its habitat or range. 2. utilization for commercial sporting, scientific, or educational purposes at levels that detrimentally affect it. 3. disease or predation. 4. absence of regulatory mechanisms adequate to prevent the decline of a species or degradation of its habitat; or 5 other natural or man-made factors affecting its continued existence (Ayensu 1981). As there has been sufficient work on threatened plants representative of a state the evaluation of threatened elements have been by and large species and site specific. The taxa have been accordingly enumerated. For example, three or fewer locations in Wisconsin State means automatically that the taxon is considered as 'endangered', even if on protected sites; In Florida, status involves conveying information as to the decrease in numbers of a species with passage of time and in Nevada subjective aspect of the plants i.e. attractive to collectors *versus* inconspicuous; etc. The categories are such as rare (R), presumed extinct (PE), endangerment (E), vigour (V) and distribution (D). Subsystem with first priority (reproductive, breeding, pollination, reproductive potential and second priority (phenology, morphology, isolating mechanisms and reproductive barriers, variation and reproductive strategies analysis.

New Zealand

Given (1981) while studying the threatened plants of the country provided a data sheet for the assessment and evaluation *vis-a-vis* conservation on the basis of 30 major sub-heads, to cite a few, important ones are: distribution (endemism, current and past distribution, type locality, range of distribution from herbarium and other records), disturbance of taxon (Past and future) conservation measures (undertaken and recommended) and conservation. 1. Status summary (Endangered number, presumed extinct in wild, Endangered), 2. Rare, 3. depleted and 4. local; Threat number with 0-4 evaluating figures covering distribution, accessibility, attractiveness, abundance change, reservation, site threat and conservation measures in operation (non-existent, minimal, adequate, good, excellent).

Australia

Threatened plants of Australia have been assessed for their status through *distribution category* (numerical code) covering type locality collection, restricted endemics, rare species in small populations, disjunct elements, taxa at risk; and conservation status (alphabetical code) like **X** for species believed to be extinct. Not collected in recent years, **E** 'endangered' **V** 'vulnerable', **N** other species listed under numerical code and not known in protected areas; and **C** species of numerical code not currently 'endangered' or 'vulnerable' and which are known to occur in national parks and other declared reserves. The country has 2053 indigenous species i.e. 8-10 per cent of the total flora listed as at risk.

Commonwealth of Independent States (CIS) of erstwhile **USSR**

The countries of erstwhile USSR (12 countries Turkemenistan,Uzbekistan, Tazikstan, Kyrgyzstan ,Kazakhastan in Asia and Estonia, Latvia, Lithuania, Russia, Belorussia,Ukraine, Moldavia in Europe) are reported to have in total 2000 species which are rare and endangered i.e. about 10 per cent of the flora (Beloussova and Denissova, 1981). The "*Red Data Book of USSR*" containing an account on 444 taxa was published in 1978. Information on each of the species is provided on scientific entity status distribution, habitats, population, reproductions, causes of decline cultivation, conservation measures taken and proposed and references. The criteria for the categorisation are assessed through 1. The size of the range and number of populations covering the identification of several groups such as (a) 'unique' - the only plant or the only population with rare individuals up to 3 (b) '*the rarest*' - one or a few populations with rare individuals totalling upto 20 (c) '*exceptionally rare*' - several population with dozens of individuals in each of them totalling upto 100 individuals '*very rare*' - upto ten populations, with the total number of individuals not more than 1000. Species occurring sporadically and rare are kept under '*sufficiently rare*' and '*rare*'. 2. Habitat conditions (deterioration in part or within the whole range), and 3. Vitality of species (a prosperous/an oppressed/vanishing species).

Sweden

In Sweden the assessment of threatened plants began under "Project Linnaeus" in mid 1970s. The studies revealed about 200 taxa as threatened of these 28 are reported to be *Extinct* (since 1850), 34 *Endangered*, 57 *Vulnerable* and nearly 50 *Rare*. (Nilsson, 1981). Categories have been assessed through field inventory, analysis of the results and practical conservation.

Federal Republic of Germany

Federal Republic of Germany has about 2667 indigenous vascular plants, out of which about 31 per cent (822 species) are threatened (Sukopp *et al.* 1978; Sukopp and Trauttmann, 1981). Red Data Book covering such plants represents 58 species '*extinct* or *presumed extinct*; 161 '*endangered*', 388 '*vulnerable*' and 215 '*potentially threatened*'. Removal of the ecotones between two land-use systems followed by drainage caused habitat elimination of most species. Out of 11 causal agents (agriculture and land improvement) = 397 spp.; tourism & recreation = 112 spp.; mining & quarrying = 106 spp; urban - industrial utilization = 99 spp; water management = 99; forestry & hunting = 84; disposal of refuse and waste water = 67; fishery = 37; military = 32; transportation = 19 and science = 7). Agriculture is regarded responsible for the decline of as many as 397 species. As many as 15 causes namely, elimination of habitats (210), drainage (173), abandonment of land use (172), land filling, grading (155), alteration of land use (123), strip mining and removal of top soil (112), mechanical impacts i.e. trampling, camping etc. (99), herbicide application (89), impacts of weeding, clearing, fire (81), dredging of rivers and lakes (69), collecting (67), water eutrophication (56), discontinuance of periodical soil disturbance (42), water pollution (31) and urbanization (20). Figures in brackets correspond to the threatened taxa.

Other European countries

In most of the European countries causal factors behind the threatened status of taxa seem to be similar. In Eastern England threat value of the species has been assessed on the basis of field survey, site-specificity covering different habitats, site histories and mapping the population with numbers of plants, associated species, conservation status and ownership (see Crompton, 1981). Perring and Farrell (1977) considered the features such as attractiveness and usefulness of plants of England which are exposed to unethical collection, accessibility of the terrain, rate of decline and percentage of localities in nature reserves for threat value assessment. The categories have been provided with index number with arbitrary values.

Noble Hardwoods Network of European Forest Genetic Resources Programme (EUFORGEN) covering different European countries assess and evaluate the threat value of forestry important species such as poplars, birch, beach etc. through field surveys and observations, genetic mapping through the advance techniques and practical conservation measures.

India

Assessment and evaluation of threat value of plant species of Indian region by and large were in past based upon the Red Data Book categories used by IUCN to indicate the degree of threat to individual species of flora and fauna in their wild habitats. Details of these categories have been separately dealt in the present treatise. With a view to providing strategic guidance for application of intensive management and information collection techniques to threatened taxa Conservation Assessment and Management Plan (CAMP) workshops were organised by Conservation Breeding Specialist Group (CBSG), Species Survival Committee of IUCN in southern and western India (Walker and Molur, 1997). The studies brought out reference manuals, biological information sheets on economically important plant groups, assignment of new IUCN categories, taxon data sheet covering the features such as distribution (historical, current), concentrated migration regions, extent of occurrence, area of occupancy, locations, populations trends, generation time, population (world, regional), data quality (general field study, informal field sightings, indirect, Herbarium/Museum, popular belief/hearsay), recent field studies, threats, categories, national and international assessment for Red Data Book, recommendation for research management and population and habitat viability Assessment (PHVA).

The IUCN Threatened Plants Committee enlist tentatively as many as 23 threats to plant ranging from industrialization and urbanization to pressure from introduced plants and natural causes. There are certain gaps in knowledge on the factors behind the varying degree of threats with respect to speceis and sites, for example, species diversity, seismic incidence, loss of habitat, biological, pollination, human interference, invasion by weeds, trade, disease, genetic problem, environmental pollution, subjective (attractiveness), accessibility etc. In Indian context the factors and parameters for threat value index are due to the fact that the threatened elements fall under different biotically and edaphically influenced zones. Moreover, there are different floristically rich regions which remain still un/ or under explored. Jain and Sastry (1981) believed that the existence of only a few herbaria having pre-1900 collections posses limitations in assessing the states of habitats or areas of past distribution of threatened flora. Jain (1981) reviewed briefly the various qualitative categories for the assessment of threats as operative on the flora of California and Britain, emphasizing on the distribution, decline of species on a 10 sq.km. grid (0 if decline is less than 33 per cent, 1 if less than 33-66 per cent, 2 if more than 66 per cent) and one sq. km. is considered as one locality for a species. Threat numbers 0-4 are given depending on 16, 10-15 (1), 6-9 (2), 3-5 (3) or less than 3 localities (4) of occurrence, respectively. A range 0-2 was made for subjective assessment such as species not attractive (O), moderately (1) or very attractive (2). Evaluation of conservation status was made on the basis of the occurrence of taxa in protected or in vulnerable habitats. If more than 66 per cent of its localities fall in nature reserves, it is 0 and if less than 33 per cent of its localities are in reserves the threat number is 3. Category of 'Endangered' is subjected to the rate of decline (0, 1, 2 = 33-66 per cent old and new locality), localities of presence (0,1,2,3,4 = 16-1 per cent) attractiveness (0, 1, 2 = ornamental), conservation (localities in reserves: 0, 1, 2, 3, = 66-33 per cent), remoteness (0, 1, 2 = island) and accessibility (0, 1, 2 = cliff). Jain and Rao (1983) compiled the contributions presented in a symposium during 1981 by different workers on the status, survey, range extensions and conservations of threatened plants of India ranging from higher plants to liverworts and lichens.

In India the problem of threatened plants was first discussed in the 11th Technical Meeting of the IUCN in 1969 and in Forestry Conference (Sahni, 1970, 1973). For studies on threatened plants of forest origin attention began focussing with the publications on the subject by several workers mainly dealing with the reports on the occurrence and decline of rare forest genetic resource. On general flora, systematically works began with the launching of Threatned Plant Project: *Project on study, survey and conservation of endangered species of flora* (POSSCEF) by Botanical Survey of India, Howrah. The project activities resulted into publications of Red Data Book on Indian Plants into three volumes (Sastry and Nayar, 1987-1990) dealing with only 285 forest species, whereas *ca* 300 forest taxa are required to be added.

Biswas and Keerti (*in editus*) based upon the source of information from literature, herbarium and field assessed and evaluated forest species. Biogeographic (10 parameters) and phytogeographic entity (32), ecological preference (33), area of occupancy, biology, habit with morphological and phenological features of conservation value, pollination mechanism, carpology, ethnobiology etc. were significant parameters to be considered. Threat value assessment is to be made from *factorial* parameters like biological, physical, pathological (0-4), *distributional/population/ ethnobiological* (Extent, Estimates, Reduction, Location, Remoteness, Attractiveness, Ethnobiology, Extinction probability) 0-4, and conservation measures (Index, *in situ* and *ex situ*, Rehabilitation, Relocation/shifting the site) in figures ranging from 0-4 as provided in copy of Data Sheet provided in the present work. Threat Value Evaluation is carried out by assigning and considering the latest IUCN categories, such as Critically Endangered (CR) with a maximum of figures of 67, Endangered (EN) 48, Vulnerable (VU) 44, Threatened (T) 44 and 20-67 covering different categories such as Rare, Endangered and Vulnerable; and Rare (R) 20-30. The general criteria and indicators followed are primarily based upon IUCN categories.

Database sheets on some important species are provided here as a reference and case studies. Only applicable features are highlighted for the species assessed and evaluated for threat and conservation values.

DATA FORMAT

Region of Studies

I Scientific Name, Family, Vernacular Name

II Species diversity: World/India/Region

III Habitat: Forest Type/Elevation/Associates/Area of occupancy.

IV: Distributional Range: India & Regional

V: Phytogeographical/Morphological entity.

VI: Threat Value Assessment [**A**: Factorial/Biological/Competition]; **B**: Distributional - Extent/Estimates/Reduction/Location/Remoteness/ Subjective assessment/Ethnobiology/Extinction probability; Conservation measures (operative)].

VII: Threat Value Evaluation

VIII: Status

IX: Further Research Works

X: Remarks

Example: 1

Data Base Sheet for Threat & Conservation Value Assessment & Evaluation (Only applicable features are entered)

Assessment and Evaluation for studies on the conservation of plant species of India.

Region of studies: *Vicinity of Dholkhand, Siwaliks, Rajaji National Park, Uttaranchal.*

I. **Scientific Name**: *Phoenix acaulis* Buch. - Ham. ex Roxb. (*Arecaceae*)

Family: *Arecaceae.*

Vernacular Name: *Dwarf Palm, Khajuri, Jangli khajur.*

II. **Species Diversity**: *World Over*: **15**: *India*: **7**: *Region of studies*: **2.**

III. **Habitat**: Forest type: *Tropical dry deciduous*: Elevation: *ca. 1000 m.* Associates (Main): *Ziziphus mauritiana* var. *fructicosa, Imperata cylindrica, Terminalia tomentosa*. Area of occupancy: *less than 5 sq. km.*

IV. **Distributional Range**: *India and Regional*

V. **Phytogeographical/Morphological entity**: Individuals scattered.

VI. **Threat Value assessment**

A: Factorial: *Meteorological*: *Fire* (4). *Biological*: *Genetic Problem* (1) - Dioecious nature. *Pollinators* (3). *Unisexuality* (4): Female less in number compared to male (1:15). *Germination interdependent* (2): Seeds growing under *Ziziphus* bushes surviving better other wise devoured by wildlife.

Competition: *Associates pre-dominant* (3). *Biotic*: *Grazing* (2) - Cattle herds of Gujjars graze and trample the seedlings. *Fire* (5) Periodically burnt.

Critically Endangered: 24.

CR: 29; **EN**: 20; **VU**: 20; **T**: 6-29; R:6.

B: Distributional *Extent*: Small geographic range (3). *Estimates*: Few but biotically disturbed (4). *Reduction*: More than 66 per cent (2). *Location*: Localities 1-2 (4): *Remoteness*: Locality accessible easily (2). *Subjective assessment*. Plant moderately attractive (1) because of its foliage and fructification. *Ethnobiology*: Lesser known (1). *Extinction probability*: Survival is possible by checking biotic interference (2) and by translocation/ reintroduction etc. (3).

Critically Endangered (CR) : **22**

CR: 25; **EN**: 20; **VU**: 12; **T**: 6-25; **R**: 6-9.

Conservation Measures (Operative).

Index: Less than 33 percent in nature reserves but under threat (3). *In situ*: Conservation forestry (1) *Ex situ*: Genetic and in sanctuary (1). *Ex situ*: manipulation (2) Tissue culture (2). *Relocation/Shifting the site* (4).

Critically Endangered: 13.

CR: 13; EN = 12; VU: 12; T: 8-13; **R**: 8.

VII. **Threat Value Evaluation**

Phoenix acaulis occurring in the region of its distributed site specifically '*Critically Endangered*' (24 + 25 + 13 = 62) with figures approaching 67, being the maximum figure of the category. The species appears to have moved to the present category from '*Endangered*' during the recent past due to the threats assesed here.

VIII. **Status**: Endangered to Critically Endangered.

IX. **Further Research Works**: Status survey, Monitoring, Habitat management and Life history studies.

X. **Remarks**: There is need to study the life history of this spectacular palm with particular emphasis on reproductive biology, phenology and its interdependence for seed germination. Localities of the occurrence of this species in other parts of the country are required to be monitored and managed.

*Example:*2

Region of studies: Doon Valley and adjoining regions, Uttaranchal.

I. **Scientific Name**: ***Dalbergia lanceolaria*** Linn.f.

Family: *Fabaceae*

Vernacular name: *Takoli, bithua* (Hindi).

II. **Species Diversity**: *World over*: **100** *India*: **27**: *Region of studies*: 2

III. **Habitat**: Forest type: Tropical dry deciduous.

Elevation: Upto 1500 m

Associates (Main): *Albizia procera, A. lebbeck, Lannea coromandelica, Shorea ro busta, Terminalia tomentosa, Erythrina suberosa, Butea monosperma* etc.

Area of occupancy: *less than 5 sq. km.*

IV. **Distribution Range**: *India and Regional*

V. **Phytogeographical/Morphological entity**: *Individuals scattered*

VI. **Threat Value Assessment**:

A. Factorial: **Biological**: Genetic problem (1). **Competition from associates**: *Without immediate impact* (1). **Biotic**: *Constructions* (3), *Fire* (5).

Threatened (T): 10

CR = 29; **EN**: 20; **VU**: 20; **T**: 6-29; **R**: 6.

B. Distributional: *Extent*: Found in small number but occurring in several areas (2). *Estimates*: Individuals few and healthy but reproductively poor (1) and biotically disturbed (4). *Reduction* 33-66 per cent (1) *Location* 3-5 (3). *Remoteness*: Locality moderately accessible (1).

Subjective assessment: Tree moderately attractive for exploitation (1). *Ethnobiology* lesser known (1). *Extinction probability*: To be saved by checking biotic interference (2) and *ex situ* measures (1).

Vulnerable to Endangered: 17.

CR: 25; **EN**: 20; **VU**: 12; **T**: 6-25; **R**: 6-9.

Conservation Measures: *Index* Less than 33 percent in nature reserves but under threat (3). *In situ*: Conservation forests (1). *Ex situ*: Botanic Garden (2). Gene banks (2) and Genetic manipulation (2). *Tissue Culture* (2).

Vulnerable: 12

CR: 13; **EN**: 12; **VU**: 12; **T**: 8-13; **R**: 8.

VII. **Threat Value Evaluation**: *Dalbergia lanceolaria* in its region of occurrence is '*Vulnerable*' to several threats. Threat value assessment reveals that the taxon is gradually entering into '*Endangered*' category if the factors influencing the tree are not checked immediately. Figures deduced through evaluation put the species under '*Threatened*' category but the figures on the distributional extent, population estimates, decline of the habitat, subjective assessment, conservation measures operative and proposed consider it as '*Vulnerable*'.

VIII. **Status**: Vulnerable

IX **Further Research Works**: Status survey in other areas, Monitoring, Habitat management, Life history.

X. **Remarks**: Tree occurs occasionally in the Siwalik and sub-Himalayan tracts. There is no recent collection of the specimen of the species. Flowers in May-June and fruits during September and October. The species is closer to *D. sericea* G. Don but differs by larger panicles and lesser number of leaflets (11-25). Field surveys in Siwalik tracts (Chirak Block and Jhabri Sot in Mayapur Block) and Raiwala forests are required be explored.

Example 3:

Assessment and Evaluation* of *Threatened Species of Garhwal and Kumaon Himalaya, Uttaranchal, India.

Region of Studies: *Mussoorie, Chakrata and Kumaon hills, Uttaranchal.*

I. Name:

Scientific: ***Trachycarpus takil*** Becc.

Family: ***Arecaceae***

Vernacular/Common/English: ***Thakil***

II. Species Diversity: World Over *8*; India *2*; Region of Studies *1* ***(wild).***

III. Habitat: ***Thakil (Palm) Mt., Eastern Kumaon (Sarju - Kali Rivers), Uttaranchal.***

Forest type: ***Temperate Forest type.***

Elevation: ***1600-2200 m***

Associates (Main): ***Quercus leucotrichophora, Rhododendron arboreum etc.***

Area of occupancy: ***less than 5 km*** ^{2}, 6-300 km^2

301-1000 km^2

IV. Distributional Range:

General (global), ***India (National), Regional (State) Local***

V. Phytogeographical/Morphological entity:

Continuous/Disjunct. ***Endemic/*** Relict/***Individual Scattered/***Monotypic /With sympodial clumps/With monopodial clumps

VI. Threat Value Assessment:

(A) Factorial: ***Meteorological:*** Rainfall (0) Relative Humidity (1) Temperature (2) **Snow/Avalanche (3)** ***Physical Calamity:*** Flood (0) **Drought (1)** Seismic (2) Valcano (3) Fire (4) ***Biological***: **Poor coppicing (1) Poor modification (2)** Poor root system (3) **Genetic Problem (1)** Gregarious flowering (2) Pollinators (3) Unisexuality (4) Sterility (5) Seed setting (1) **Germination interdependent (2)** Viability (3) ***Disease/Pest:*** E conomic threshold (0) E conomic injury level (1) ***Mortality beyond revival (2)*** **Competition:** Main species phenologically sound (0) ***Admixture without immediate impact (1)*** A ssociates gregarious periodically (2) Associates predominant (3) Associates wiped out endemic (4) ***Biotic:*** Tourism (0) Fuel (1) Grazing (2) ***Trade (3)*** Constructions (3) ***NWFP (4)*** Shifting cultivation (4) Warfare (4) Fire (5)

Endangered: 20.

CR = 29; **EN**: 20; **VU**: 20; **T**: 6-29; **R**: 6

(B) Distributional/Population/Ethnobiological:

Extent: Distributed largely (0) Restricted to a very few localities/single locality but in large no. (1) ***Found in small no. but occurring in several areas (2) Small geographic range (3)***

Estimates: Individuals few healthy (wild) (0) ***Individuals few healthy but reproductively poor (Wild) (1)*** Individuals few but healthy (cult.) (2) Individuals few, healthy but reproductively poor (cult.) (3) ***Few but biotically disturbed (4).***

Reduction: (per decade) Decline < 33% (0). 33-66% (1). ***> 66% (2)***

Location: Over 16 No. localities (0). 10-15 (1). ***6-9 (2)*** 3-5 (3). 1-2 (4)

Remoteness: Locality not easily accessible (0), Moderate (1), ***Easily (2)***, Locality known but not easily accessible (0), Moderate (1), ***Easily (2)***.

Subjective assessment: Plant attractive (0), Moderate (1), ***High (2).***

Ethnobiology: Uses not known (0), Lesser known (1), ***Well used (2).***

Extinction probability: Can be saved by *in situ* measures (0) ***By ex situ* (1), *By checking biotic interference* (2), *By translocation, reintroduction etc.* (3)**, cannot be saved (4).

Critically endangered: 28

CR: 25; **EN**: 20; **VU**: 12; **T**: 6-25; **R**: 6-9

(C) Conservation Measures: (Operative and Proposed)

Index: > 66% in nature reserves (0), > 33-66% in nature reserves (1), 33% in nature reserves (2), ***< 33% but under threat* (3),** Well preserve (0) .

In situ: ***Conservation Forests* (1).** Sacred groves (1), Sanctuaries (1 Preservation Plots (1).

Ex situ: Plantation Forests (2), Arboreta (2), BG (2), Gene Banks (2 Seed Banks (2), Cryopreservation (2). Genetic Manipulation (2) Tissu Culture (2). ***Measures not known* (3), *Education* (3)**

***Rehabilitation/Reintroduction* (3).**

Relocation/Shifting the site (4).

***Critically endangered* 17.**

CR: 13; **EN**: 12; **VU**: 12; **T**: 8-13; **R**: 8.

VII Threat Value Evaluation

Categorisation (* New IUCN categories)			**Factorial + Distribution-cum-Population trend + Conservation measures (Operative)**
CR	=	*Critically Endangered:	29+25+13 = 67
EN	=	Endangered:	20+16+12 = 48
VU	=	Vulnerable:	20+12+12 = 44
T	=	Threatened:	6-29+16/25/12/6 to 9+12/13/12/ 8 to 9 (CR/EN/VU/R)=20-67
R	=	Rare:	6+6 to 9+8 = 20-30

[Extinct (Ex); *Extinct in wild (Ew); *Low Risk (LR); *Conservation Dependent (CD); Indeterminate (I); *Data Deficient (DD); *Not Evaluated (NE); Least Concern (LC); Out of Danger (0) & Reproduction (REP)]

VIII Status: Extinct/Extinct in wild/***Critically Endangered/ Endangered/*** Vulnerable/Threatened/Rare/Low Risk/Conservation Dependent/ Indeterminate/Data Deficient/Not Evaluated/Least Concern/Out of Danger/ Reproduction.

IX Further Research Works:

Status Survey*: *Monitoring, Limiting Factor Research, Limiting Factor Mangement, ***Habitat Management,*** Taxonomic studies, ***Life history studies.***

X Remarks: *Trachycarpus takil Becc., an endemic palm of Garhwal and Kumaon* (Uttaranchal) *has its historical distribution in 'Takil Mt.' of Kumaon hills where the species has been indiscriminately cut by local inhabitants for fibre. The wild population has declined considerably. In certain threatened localities of Mussoorie and Chakrata hills where this palm has become naturalized, it appears to be under threat once the use of the plant is known to the local inhabitants and thus may meet the similar fate as in Mt. Takil in Kumaon. The conservation of this palm has become an international concern now. This concern is highlightened by Dr. Gibbons of 'The Palm Centre, London in 1991 with a mention that the local people living near the locality of the occurrence of the species should be made aware about how this tree grows, its rarity and value, and the desirability of putting an end to the senseless cutting of the young adult plants before they have had a chance to produce seeds. Without this, the tree is undoubtedly doomed to extinction.*

The palm is 'Critically Endangered' as there is sharp decline of the species from its natural area of occurrence during the past 6-7 decades.

Example 4:

Assessment and Evaluation of Threatened plant Species of Uttaranchal.

Region of Studies: ***Doon Valley and adjacent regions,Uttaranchal..***

I. Name

Scientific: ***Sophora mollis Grahm ex (Royle)Baker***

Family: ***Fabaceae***

Vernacular/Common/English: ***Sakina (Garhwal), a name also applied to all Indigoferas, which it somewhat resembles in foliage.***

II. Species Diversity: ***World Over 50, India 6, Region of Studies 1.***

III. Habitat

Forest type: ***Tropical Mixed Deciduous Forests***

Elevation: ***500-1500 m.***

Associates (Main): ***Colebrookia oppositifolia, Rubus spp., Strobilanthes spp., Debregeasia spp., Villebrunea spp. and weeds.***

Area of occupancy: *less than 5 km²* 6-300 km²

301-1000 km²

IV. Distributional Range:

General (global)/ ***India (National)/Regional (State)/Local.***

V. Phytogeographical/Morphological entity:

Continuous/Disjunct/Endemic/Relict/***Individual Scattered/***Monotypic With sympodial clumps/With monopodial clumps.

VI. Threat Value Assessment:

(A) Factorial: ***Meteorological:*** Rainfall (0), Relative Humidity (1), Temperature (2), Snow/Avalanche (3), ***Physical Calamity:*** Flood (0), ***Drought* (1),** Seismic (2), Valcano (3), ***Fire*** (4). ***Biological:*** Poor coppicing (1), Poor modification (2), Poor root system (3), ***Genetic Problem* (1)**, Gregarious flowering (2), Pollinators (3), Unisexuality (4), Sterility (5), Seed setting (1), Germination interdependent (2). ***Viability* (3).** ***Disease/Pest:*** Economic threshold (0), ***Economic injury level* (1).** Mortality beyond revival (2). **Competition:** Main species phenologically sound (0). Admixture without immediate impact (1). ***Associates gregarious periodically* (2).** Associates predominant (3). ***Associates wiped out endemic (4).*** ***Biotic:*** Tourism (0), Fuel (1), Grazing (2), Trade (3), Constructions (3), Shifting cultivation (4), Warfare (4), Fire (5)

***Threatened approaching Endangered category* :17**

CR = 29; **EN**: 20; **VU**: 20; **T**: 6-29; **R**: 6

(B) Distributional/Population/Ethnobiological:

Extent: Distributed largely (0), Restricted to a very few localities/single locality but in large no. (1). ***Found in small no. but occurring in several areas (2)***, Small geographic range (3).

Estimates: Individuals few healthy (wild) (0), ***Individuals few healthy but reproductively poor (Wild) (1),*** Individuals few but healthy (cult.) (2) Individuals few, healthy but reproductively poor (cult.) (3). ***Few but biotically disturbed (4).***

Reduction: (per decade) Decline < 33% (0), ***33-66% (1)***, > 66% (2).

Location: Over 16 No. localities (0), 10-15 (1), ***6-9 (2),*** 3-5 (3), 1-2 (4).

Remoteness: Locality not easily accessible (0), Moderate (1), Easily (2). Locality known but not easily accessible (0), Moderate (1), ***Easily (2).***

Subjective assessment: Plant attractive (0), Moderate (1), ***High (2).***

Ethnobiology: Uses not known (0), ***Lesser known (1),*** Well used (2).

Extinction probability: Can be saved by *in situ* measures (0), ***By ex situ (1),*** By checking biotic interference (2), By translocation/reintroduction etc. (3), cannot be saved (4).

Endangered :18

CR: 25; **EN**: 20; **VU**: 12; **T**: 6-25; **R**: 6-9

(C) Conservation Measures: (Operative and Proposed)

Index: > 66% in nature reserves (0), > 33-66% in nature reserves (1), < 33% in nature reserves (2), ***< 33% but under threat (3)***, Well preserved (0)

In situ: ***Conservation Forests (1),*** Sacred groves (1), Sanctuaries (1) Preservation Plots (1).

Ex situ: Plantation Forests (2), Arboreta (2), ***BG (2)*** Gene Banks (2), Seed Banks (2), Cryopreservation (2), Genetic Manipulation (2), ***Tissue Culture (2).***

Rehabilitation/Reintroduction (3)

Relocation/Shifting the site (4)

Critically Endangered :13

CR: 13; **EN**: 12; **VU**: 12; **T**: 8-13; **R**: 8.

VII Threat Value Evaluation

Categorisation (*New IUCN categories)		Factorial + Distribution-cum-Population trend + Conservation measures (Operative)
CR =	*Critically Endangered:	29+25+13=67
EN =	Endangered:	20+16+12=48
VU =	Vulnerable:	20+12+12=44
T =	Threatened:	6-29+16/25/12/6 to 9+12/13/12/ 8 to 9 (CR/EN/VU/R)=20-67
R =	Rare:	6+6 to 9+8=20-30

[Extinct (Ex); *Extinct in wild (Ew); *Low Risk (LR); *Conservation Dependent (CD); Indeterminate (I); *Data Deficient (DD); *Not Evaluated (NE); Least Concern (LC); Out of Danger (0) & Reproduction (REP)]

VIII Status: Extinct/Extinct in wild/***Critically Endangered***/ Endangered/ Vulnerable/Threatened/Rare/Low Risk/Conservation Dependent/ Indeterminate/Data Deficient/Not Evaluated/Least Concern/Out of Danger/ Reproduction.

IX Further Research Works:

Status Survey, ***Monitoring,*** Limiting Factor Research, Limiting Factor Mangement, ***Habitat Management,*** Taxonomic studies, ***Life history studies.***

X Remarks: *Although the species has been reported to occur in different parts of Himalaya from Pakistan to Central Nepal between 1200-2000 m on exposed aspect in dry valleys and rocky situations but the type locality of the species is Sahastradhara valley in Indian sub-Himalayan foot hills. The type locality of the species is threatened due to severe biotic impact. The population of the taxon has declined during the past 3 decades. Reproductive biology of the species is not known. Plant being highly ornamental when in flower during February to April, and its deep foliage with pinnate leaves make the species an* *xcellent choice to plant enthusiasts and horticulturists for trade and andscaping.*

Literature Cited

Ayensu, E.S. (1978): The US Red Data Book. *Garden* **2** (5): 2-3.

Ayensu, E.S. (1981): Assessment of threatened plant species in the United States. *In*: Synge H (Ed.). *The biological aspects of rare plant conservation*. 19-50. John Wiley & Sons Ltd.

Beloussova, L. and Denissova, L. (1981): The USSR Red Data Book and its compilation. *In*: Synge, H. (Ed.). *The biological aspects of rare plant conservation*: 93-99. John Wiley & Sons Ltd.

Cook, CDK (1980): The status of some Indian endemic plants. Threatened plant *Committee News letter* 6:17-18 Kew. England.

Crompton, G. (1981): Surveying rare plants in Eastern England. *In*: Synge H. (Ed.) *The biological aspects of rare plant conservation*: 117-124. John Wiley and Sons Ltd.

Fay, JJ (1981): The endangered species program and plant reserves in the United States. *In*: Synge, H. (Ed.). *The biological aspects of rare plant conservation*. John Wiley and Sons Ltd.: 447-452.

Fosberg, FR and Herbst, D. (1975): Rare and endangered species of Hawaiian vascular plants. *Allertonia* **1** (1).

Gadgil, M. and Vartak, VD (1975): Sacred groves of India - a plea for continuing conservation. *J. Bombay Nat. Hist. Soc.* 72:314-20.

Given, DR (1981): Threatened plants of New Zealand: Documentation in a series of islands *In*: Synge, H. (Ed.) *The biological aspect of rare plant conservation*: 67-79. John Wiley & Sons Ltd.

Good, RB and Lavarack, PS (1981): The status of Australian plants at risk. *In*: Synge, H. (Ed.) *The biological aspect of rare plant conservation*: 81-91. JOhn Wiley and Sons Ltd.

Hall, AV, de Winter, M., de Winter, B. and Van Oosterhout, SAM (1980): *Threatened plants of Southern Africa*. South African National Scientific Programmes Report No. 45.

Henry, AN, Vivekananthan, K. and Nair, NC (1978): Rare and threatened flowering plants of south India. *J. Bombay Nat. Hist Soc.* 75:684-97.

Jain, SK (1981): An sight into the problem of threatened plants, *INSA News letter*, No. 66: 1-3. Indian National Sci. Academy New Delhi.

Jain, SK and Sastry, ARK (1981): *Threatened plants of India - A State of the Art Report*, Botanical Survey of India, Howrah - Techniques and constraints in survey and conservation of threatened plants and habitats in India. *In*: Synge, H. (Ed.) *The Biological Aspects of Rare Plant Conservation*: 59-66. John Wiley & Sons Ltd.

Jain, S.K and Rao, R.R. (Eds): An assessment of threatened plants of India.Proceed.of the seminar held at Dehradun,14-17 September, 1981, Botanical Survey of India, Botanic Garden, Howrah.

Leon (1980): Endangered plants in a Draft community. List of Threatened Species of Wild Flora and Vertebrate Fauna Vol.I. Prepared for the Environment and Consumer Protection Service of the Commission of the European Communities by the Nature Conservancy Council of the Great Britain, Nov. 1980.

Lucas, G. and Synge, H. (1978): The IUCN Plant Red Data Book, Morges, Switzerland.

Lucas, G. and Synge, H. (1981): The assessment and conservation of threatened plants around the world. *In*: Synge, H. (Ed.) *The biological aspects of rare plant conservation*: 3-18.

Nilsson, O. (1981): Project Linnaeus: Assessing Swedish plants threatened with extinction. *In*: Synge, H. (Ed.) *The biological aspects of rare plant conservation*: 105-124. John Wiley & Sons Ltd.

Perring, FH and Farrell, L. (1977): British Red Data Books: I Vascular Plants, Society for the Promotion of Nature Conservation, Lincoln. pp. 94.

Sahni, KC (1970): Protection of rare and endangered plants in Indian flora. *Proc. IUCN 11th Technical Meeting* 11 95-102.

Sahni, KC (1973): Protection of endemic & relict taxa in Indian Flora. *Proc. 12th Forestry Conference* Dehra Dun.

Sastry, ARK and Nayar, MP (Eds.) (1987-90): Red Data Book on Indian Plants. Volume I-III. Botanical Survey of Indian Howrah.

Sukopp, H, Trautmann, W. and Korneck, D. (1978): Auswertung der Roten Liste gefahradeter Farn - und Blutenpflanzen in der Bundesrepublik Deutsch and fur den Arten - und Biotopschutz. *Schriflenreihe fur vegetationskunde* 12: Bonn - Bad Godsberg.

Sukopp, H. and Trautmann, W. (1981): Causes of the decline of threatened plants in the Federal Republic of Germany. *In*: Synge, H. (Ed.) *The biological aspect of rare plant conservation*: 113-116. John Wiley & Sons Ltd.

Shimizu, T. and Satomi, N. 1976): A preliminary list of rare and critical vascular plants of Japan (1). *J. Fac. Arts, Shishu University, Nat. Sci.* 10:3-16.

Siddall, JL; Chambers, KL and Wagner, DH (1979): Rare, Threatened and Endangered Vascular Plants in Oreganan Interim report. Natural Areas Preserves Advisory committee, Salem.

Walker, S. and Molur, S. (1997): Conservation Assessment and Management Plan (CAMP) workshop process, Reference Manual. Second Indian edition (Rvs for *Indian* BCPP. Endangered Species Prioritisation Project. CBSG, SSC, IUCN.

Ward, DB (Ed.) (1979): Rare and Endangered Biota of Florida, Vol. 5: Plants. University Presses of Florida, Gaines ville, USA.

West, RG (1981): Preface *In*: Synge, H (Ed.). *The biological aspects of rare plant conservation*. John Wiley & Sons. Ltd.

Whitson, PD and Massey, JR (1981): Information systems for use in studying the population status of threatened and endangered plants, in *Rare Plant Conservation: Geographical Data Organisation* (Eds. L E Morse & M S Henifun): 217-236. New York Botanic Garden.

Chapter VII

Indian Region - the Reservoir of Medicinal Flora and the Conservation Needs

Indian region has a 5000-4000 B.C. years of historical past of using plant resource for myriad of diseases. As for the cultivation of plants of great significance to mankind it could be traced back during the late Mesolithic to early Neolithic age (10,000 years BC), principally for the items of food to sustain life. 80,000 species out of nearly 2,50,000 seed bearing plants occurring globally bear medicinal and aromatic values. SE Asia, Africa, Latin America and eastern Europe provide nearly 90 percent of raw materials for medicinal exploitation. It is revealed that there are actually 121 plant donated prescription drugs (Farnsworth and Soejarto, 1988) as a result of the study of about 35,000 species of plants. Out of a very rich diversity among the medicinal and aromatic flora about 5000 species are considered to have been examined thoroughly for the active principles. It is reported by Myers (1983) that only 41 species generate commercial scales to the tune of 40 billion per US dollar per year. Non formulation drugs as adjuncts are known among between 50-60 billions US dollar.

Indian region endowed with nearly 20,000 species of plants highlight 75 major species of medicinal flora out of which at least 25 were most sought after during the last decade. Federal Republic of Germany alone imported worth of 40,000 tonnes of plant dry material from India during 1980-90. Biswas *et al.* (2000) mentioned that out of about 80,000 tonnes of medicinal plants imported by western countries, India tops the list of exporters for USA and Europe with a share of over 10,000 tonnes.

India and neighbouring Nepal and Bhutan are amongst the most important medicinal plant resources collection centre. Over 500 million people receive the benefits of traditional knowledge of well-documented and standardized systems of medicines including siddha, unani, ayurveda etc. There are 460,000 practitioners of the traditional systems of medicines as per an estimate (*Hindustan Times* 27th April 1999); out of this 271,000 are registered and 7843 licensed manufactures of traditional drugs exist in India.

Medicinal and aromatic flora of India is widely distributed in different parts of the region. Such floristic resource has specific phytogeographical ıd biogeographical entities. Several zones still remain unexplored and not aving inventorized. Such areas fall under various endemic centres of plant iversity in Himalayan, western and eastern Ghats, occeanic islands and ortheastern India. As compared among the South Asian countries nearly 600 species of documented medicinal flora 540 find major use as herbal rugs (about 200 of these are used in bulk quantities having commercial otential).

pecies Richness of Medicinal Flora of Important Regions/Countries f the World

egion/Countries	Medicinal Flora (species)
uth Asia	
dia	3000
Himalayan & NE Region	250 (1500)
Western Ghats	500 (1000)
Oceanic Islands	250 (400)
Gangetic Plains	100
hutan	300
angladesh	250
epal	400 (1600)
ıkistan	400
i Lanka	300 (1500)
ast Asia	
R. China	5000
orea	746
E Asia	
alaysia	1230
donesia	500
urope	
aly	148
ortugal	89
ountries of Erstwhile USSR	228
frica	
ıst Africa	>1000
ewanda	100
ınzania	200
hiopia	365
uth America	
W Amazon Region	1300
SA	
WUSA	1500

ource: Arora & Engels 1993; Biswas *et al.* 2000)

Among the floristically rich regions of different countries economic potential of medicinal flora of PR China (5000 species), India (3000), Brazi (NW Amazon - 1300), South West USA - 1500), Malaysia (1230), Eas Africa (>1000) is well known.

During the past over 50 years numerous species of medicinal value have been discovered, invented, documented, screened and commercialized throughout the Indian region. Although explorations were made in *terra incognitio* but the attention to the medicinal properties of species have no been well studied. This has been largely due to lack of information from the dependents on such species, besides need for scientific scrutiny and screening or efficacy of the principles. The species that gathered application of newe methods of inventorization, characterization and multiplication enjoyed maximum attention for sustainable utilization. Medicinal and aromatic flora were improved through *ex situ* and *in vitro* measures of conservation. In drug industry following plants with active component mentioned in parenthesis were notable. For example.

- *Adhatoda vasica* (Vasicine)
- *Atropa belladona* (Atropine)
- *Azadirachta indica* (Azadirachtin)
- *Catharanthus roseus* (Vincristine, Vinblastine)
- Solanum spp. (Solasodine)

Cinchona spp. (Quinine): Several species of this genus are represented in India as cultigens particularly in NE India, Nilgiris in South and Satpura hills in Central India, for example *C. calioaya, C. ledgeriana, C. officinali* and *C. succirubra*.

Dioscorea spp. (Diosgenin): Among the several species found in th country *D. deltoidea* and *D. prazeri* possess appreciable quantity o diosgenin, the discovery of which was made in 1960s.

Other species are ginseng, ipecac, digitalis, liquorice, psyllium, senna mishmi teeta, *Taxus* etc.

While the attention has to be focussed on the wild relatives of well known to over exploited species of medicinal flora there is an emergent need for the conservation of rare and threatened germ plasm of various species of useful elements. Certain plant groups as specified by the families have been identified to possess active principles and analogs of compounds in development. These families as per Myers' work (1983) on anticancer drug yielding plants are as following:

Apocynaceae
Celastraceae
Compositae (Asteraceae)
Simaroubaceae
Rutaceae
Thymelaeaceae
Magnoliaceae
Annonaceae
Boraginaceae
Leguminosae (*Fabaceae*)
Liliaceae
Rubiaceae
Gymnosperms

Studies reveal that at least 70 percent of the 3000 plant species worldwide possess anticancer properties. Region wise 4.8 percent of Indian flora has anticancer properties (Myers, 1983).

The levels of exploitation of medicinal flora have reached all time high. This has resulted into decline of potential species and infraspecific forms, for example, to produce one kilogram of vincristine, an anticancer lymphocytic leukema) drug as much as 530 m tons of plant material of rosy periwinkle (*Catharanthus roseus*) is needed to be exploited. One kilogram of product costs over 1 million US dollar. *Taxus baccata* providing anticancer 10-deacity baccation III (DAB) from its needles and bark sells between 0.1-0.13 million US dollars per Kg, whereas present demand exceeds 100 million US dollars per year (Biswas *et al.* 2000). The extraction of bark causes the death of the tree. Similar are the cases with other medicinal herbs where whole of plants are removed indiscriminately before those could reach to the phenological maturity and vegetative regenerating capacity.

Several of medicinal plants are distributed throughout the country.

Botanical-cum-phytogeographical regions of Indian Medicinal Flora

Bioclimatically and phytogeographically the medicinal flora of India has occupied specific micro-habitats. The area and extent of occupancy

Widely Distributed Important Medicinal Plants

S.N.	Species	Vernacular Names	Commercial use	Parts
1.	Acorus calamus	Vacha, Bos ghur-vacha	In, Pe, Fo	Rhizome
2.	*Adhatoda vasica*	Vasak, adusa	-do-	Aerial part
3.	*Aloe barbedensis*	Kumari, ghee Kaul	In, Mo, Cos	Leaf mucilag
4.	*Aristolochia bracteata*	Ishwari, visanika	In(A), Fi	Whole plant
5.	*A. indica*	-do-	-do-	-do-
6.	*Asparagus racemosus*	Satawari	In, Fo	Roots
7.	*Bacopa monniera*	Brahmi	In, Mo, Fo	Whole plant
8.	*Chlorophytum arundinaceum*	Safed musli	In, Fo	Tuberous roo
9.	*Curcilago orchioides*	Kali musli	In, For	-do-
10.	*Gloriosa superba*	Kalihari	In(A), Fo,Mo	Roots & seed
11.	*Hemidesmus indicus*	anantamul	In,For	Roots
12.	*Plumbago zeylanica*	Cheeta lakri, Chitrak	In,Fo	Rotos
13.	*Pueraria tuberosa*	Vidari Kanda	In(A), Fo	Roots
14.	*Rauwolfia serpentina*	Sarpagandha	In, Fo, Mo	Roots
15.	*Rubia cordifolia*	Manjitha	In, Fo, Dy	Rotos
16.	*Withania somnifera*	Ashwagandha	In, Fo	Roots

Abbreviations: In = Indigenous industries mainly Ayurveda (A); Unani (U); Fo = In Folklc Mo = Modern medicine; Pe = Perfumery; Con = Condiment; Dy = Dye color.

are largely due to various abiotic and biotic influences. In several localiti the exotics have become naturalized. The successional trends are large based upon influencing factors. Regionwise some significant elements wi vernacular in parenthesis are mentioned with a view to highlighting t species diversity of the medicinal flora.

Western Himalayas: *Aconitum heterophyllum* (Ativis, atis), *Angeli glauca* (Cora, Grandrayan, Chipi), *Atropa acuminata* (*Angur sef Sagangur*), *Bunium persicum* (Siah jeera), *Colchicum luteum* (suranj siri), *Dactylorhiza hatazirea, Orchis habernarioides* (Salam panja, Sala Hathajari), *Fritllaria roylei* and *Lilium polyphyllum* (Kakoli, Kshir-Kakol *Gentiana kurroo* (Karu), *Hedychium spicata* (Kapur Kachri, Kachu *Jurinea* spp. (*Dhup, Guggal*), *Nardostachys jatamansi* and *grandiflora* (Mansi, Jatamansi), *Onosoma bracteata* and *Macrotom* *benthami* (Gaujuban and Ratanjot), *Podophyllum emodi* and *P. hexandrum* (Ben-kakri), *Picrorhiza kurrooa* (Kutki, Karu), *Rheum australe* and other species (Dolu), *Saussurea costus, S. lappa* (Kuth, Quest-e-sirin), *Swertia chirayata* (Chirayata), *Valeriana wallichii* or *V. jatamansi* (Asarun, Tagar, Sugandhbala), *Viola odorata, V. canescens* and *V. serpens* (Banafsha, Gulbanafsha).

Eastern Himalayas

Coptis teeta (Mamira, Mishmiteeta, Golden Thread) *Gaultheria fragatissima* (Indian Wintergreen), *Panax pseudoginseng* (Tapmare), *Dioscorea prazeri*. Biswas *et al.* (2000) enumerated 250 species of medicinal plants typical of this phytogeographical region.

Gangetic and Indus Plains, Hill tops and Desert

Azadirachta indica (Neem). *Commiphora wightii* or *C. mukul* (Guggal), *Curcuma caesia* (Kalihaldi, Narak-chur), *Gymnema sylvestre* (Gurmar), *Salvadora persica* (Peeloo), *Calotropis procera, C. gigantea* (Arka, Madar), *Euphorbia neriifolia* (Sehund, Thunner), *Peganum harmala* (Harmal), *Emblica officinalis* (Aonla), *Terminalia* (Arjun), *T. bellirica* (Bahera), *T. chebula* (Harar), *Phyllanthus* complex (*P. fraternus, P. nirurii*).

Deccan

Coscinium fenestratum (Marmanjil is used in place of *Berberis* sp.), *Pterocarpus santalinus* (Lalchandan - red chips of stem used in Ayurvedic system of medicine), *Strychnos nuxvomica* and *S. potatorum. Trichopus zeylanicus* ssp.*travancoricus* (*Arogyapacha*) - found in Agastyar hills of Western Ghats of Kerala and Tamil Nadu has anti-fatigue, immuno-enhancing and adaptogenic property.

Several of the potential species of medicinal value have been patented for their active principles in some developed countries. Some of such species are *Phyllanthus niruri* for Hepatitis A to Z, *Emblica officinalis* (Aids, Flu etc.), etc.

Conservation Needs and Strategies on Medicinal and Aromatic Flora

With the impact of various influencing factors the medicinal flora of the country is on the decline in terms of biological, (physiological, phenological, regeneration), ecological (habitat - genepool, alteration, hotspots, phytogeography) and conservation aspects (*ex situ*, *in situ*, *in vitro*). With regards to this significant group of plants adequate research and development (R & D) support using recent techniques of scientific and technological advances (S & T) is required. Following provides the various strategies.

1. Inventorization and Characterization

(i) Development of baseline database on species and genetic diversity of specific sites.

(ii) Taxonomic characterization using recent trends (biochemical, DNA, RFLP, PCR etc., Cytological, Palynological tools).

(iii) Manpower (capacity building).

(iv) Funding mechanism (collaborative).

2. Monitoring and Assessment

(i) Development of cost effective methodology for quantitative assessment of specific site.

(ii) Application of remote sensing (RS) and GIS (Geographic Information System) technology in identifying critical habitats at landscape level. Conforming RS and GIS database with ground reality.

(iii) Application of GPS (Global Positioning System) for integration.

3. Ex situ Conservation

(i) Development of nursery and planting technology for threatened taxa.

(ii) Development of gene bank (field oriented) on target species.

(iii) Networking for coordination and collaboration at inter disciplinary and inter institutional levels.

(iv) Application of cost effective ways of conserving biological parts of medicinal flora (possibly through grassroot based innovations and inventions).

4. In situ Conservation

(i) Development of microhabitats, endemic centres, hotspots. A separate Working plan circle in Forest Department may be constituted.

(ii) Application of recent trends in the field of biodiversity conservation to the existing preservation plots, sacred groves and specific sites falling under 89 National Parks (NP) and 489 Wildlife sanctuaries (WS) of the country.

The states of Uttaranchal and Uttar Pradesh have 36 number of key areas of conservation with 7 NPs and 29 WS spread over 13005.36 sq. km (National Wildlife Database Cell, WII. Dehra Dun as per date 28th Feb. 2000). Several of these areas have not yet been thoroughly inventorized and species characterized for the wealth of medicinal and other economic uses.

5. Utilization

(i) Ethnobiological based innovations and inventions and integration of indigenous traditional knowledge (IK) through scientific scrutiny.

(ii) Introduction of improved methods of harvesting and optimum loss of unused but potential products considering the sustainability aspects.

While the conservation needs are being prioritized and new technique being developed to improve upon the various aspects of medicinal flora, the lacunae of other spheres of the subject are not being worked out. This mainly concerns with the rights of communities, region and the country over the intellectual property.

Indigenous knowledge systems are similar to general scientific information in that they are part of public knowledge. Intellectual Property Rights (IPRs) have so far been applied to novel and discreet intellectual goods rather than to public goods such as knowledge systems. The usual criteria for recognizing IPRs, novelty and non-obviousness, generally tend to ignore the knowledge systems of rural and tribal families, although they are often characterized by a high degree of inventiveness. While the knowledge itself may not be patentable, the product of the knowledge, namely "folk" varieties, land races and genetic diversity at the intraspecific level, provide the basic raw material for modern plant breeding and biotechnology (Swaminathan, 1994).

Since the Convention on Biological Diversity signed at the Rio Summit in 1992 only reaffirmed the sovereign rights of the governments over their plant genetic resources but remained silent on the contentious issue of proprietary control over the products developed from these resources. The multinational drug industry is using the long arm of the WTO (World Trade Organization) to maintain its global monopoly. The initiatives of WTO over the control of markets may jeopardize the existing medicinal plant based health care systems, which are traditionally followed in the country. A pro-active approach on the issue of intellectual property control over medicinal plants could protect the national interests and provide an economic handle to dictate the global market in herbal medicines as well. It is true that a dilemma prevails about how to provide patents to formulations and products, which have been developed over hundreds of years (Shah, 1997). It remains a fact that hundreds of patents are being taken out on traditional medicines the world over.

Literature Cited

Arora, RK and Engels, JMM (1993): Genetic resources in medicinal and aromatic plants. *Acta Horticulture* 330:21-38.

Biswas, Sas. (1988): Rare and threatened taxa in the forest flora of Tehri Garhwal Himalaya and the strategy for their conservation. *Indian J. Forestry*, B 11(3):233-237.

Biswas, Sas. (1990): Ethnobotanical studies on North eastern region of India with particular emphasis on conservation of biological diversity. Proceed. 2nd International Congress of etnobiology, "The Challenges of Ethnobiology in the 21st Century" held at Kumming, P.R. China, Oct. 22-26. *Symposium session* 5 (Abs.):128.

Biswas, Sas. (1991): Rare and endangered flora eastern Himalaya and the measures for its conservation. *In*: G.S. Rajwar (Ed.) *Advances in Himalayan Ecology*, 6:267-273. Yesterday, Today & Tomorrow Pub. Delhi.

Biswas Sas. (1991a): Rare and endangered plants of economic value -Coptis teeta Wall. MFP News.

Biswas, Sas., Jain, S.S. and Sumer Chandra (2000): Some rare and threatened medicinal plants of Himalayan and North eastern regions of India and strategies for this conservation. *J. Non Timber Forest Products* 7(1/2):101-111.

Farnsworth, N.R. and Soejarto, D.D. (1988): Global importance of medicinal plants. *In*: *The Conservation of Medicinal Plants* (Eds. O. Akerele *et al.*). *Proc. International Consultation,* 21-27 March, 1988, Chiang Mai, Thailand: 25-51. Cambridge University Press, Cambridge, U.K.

Myers, N. (1983): Plants as sources of anticancer drugs. A. Wealth of Wild species - storehouse for human welfare. Colorado :105.

Parhar, K. and Sas. Biswas (1998): Medicinal plant resources of Doon Valley and adjacent Siwaliks-I. *J. Non Timber Forest Products* 5(1\2):29-33.

Rawat, JK (2000): Research and development need for conservation of vegetative biodiversity in India forests. "One Week Compulsory Training Course" for IFS Officers on Biodiversity Conservation, 25-29th Sept. 2000. Lecture Notes: 1-4. Forest Research Institute, Dehra Dun.

Shah, NC. (1982): Herbal folk medicines in northern India. *J. Ethnopharmacology* 6:293-301.

Shah, NC. (1997): Conservation of wild medicinal plants: need for a comprehensive strategy. Endangered species. December 15-17 Kurukshetra.

Swaminathan, MS (1994): Inaugural address: Fourth International Congress of Ethnobiology, Lucknow, Nov. 17. 1994.

Domestic Animal Biodiversity - Breeds, Wild Origin And Threats

Agro-ecological diversity of India has contributed immensely to the domestic animal biodiversity. Species diversity has reached to the occurrence and distribution of breeds and nondescript animals due to unmindful interbreeding. Various breeds of animals such as 30 of cattle, 427 sheep, 107 buffalo, 20 goats, 87 camel, 187 poultry, 6 of horse, 3 of pig and several of mithun, ducks, donkey, turkey etc. make the diversity remarkably rich. Nearly 8 per cent of total world cattle and goat breeds and about 27 percent of buffalo are found in this country. As per a report (see Nivsarkar, 1999), the region contains nearly 16% of the total world cattle population, 56 percent of buffaloes and 17% of goats. In context to Asia the population figures are presented as buffalo (58%), cattle (46%), camel (30%), and goat (28%).

Significantly, as a result of interbreeding among the specific groups genetically oriented phenotypes have appeared which is represented by a small percentage of recognized indigenous breeds and a very large section of non-descript indigenous animals which do not match to any specific traits of breed.

Introduction of exotic breeds of cattle and sheep, to cite two examples during the last four decades appears to have resulted into more of non-descript categories.

A comparison to the number of Breeds of Livestock species

Group	World	America	Europe	Africa	India*	ICAR** 1999	Indian % to world
Sheep	920	70	42	73	59	42	6.41
Cattle	787	110	305	120	61	30	7.75
Chicken	606	45	406	57	-	18	2.97
Horse	384	65	185	31	9	6	2.34
Pig	353	52	129	13	3	3	0.85
Goat	351	23	113	34	29	20	8.26
Donkey	77	10	16	6	3	3	3.90
Buffalo	72	3	2	2	19	15	26.39
Camel	56	-	-	-	-	8	14.29

Source: * World Watch List, FAO, 1995; ** Nivsarkar (1999)

Population size and recognized breeds of Livestock species (in '000)

Group	World	India	%	% non breed
Cattle	13,14,557	2,09,084	15%	75%
Buffalo	1,64,242	92,190	40%	60%
Chicken	1,36,82,698	3,42,500		60%
Sheep	10,61,609	56,472	30%	68%
Goat	6,93,261	1,20,600	30%	65%
Horse	60,811	990	-	-
Camel	19,603	15,520	-	-
Pig	9,57,469	16,005	-	-

Source: FAO Data Bank, 1998

For various breeds of indigenous species following projection on population has been made by 2020. (*see* Nivsarkar 1999).

Group	**Population (m)***
Cattle	250
Buffalo	130
Sheep**	30
Goats	280
Camels**	0.5
Horses** & Ponies	0.5
Yak	-
Mithun	-
Poultry	1500
Others (Mules, Donkey**, Pig etc.)	10

* Based on population structures of 1982-87.

** Would register a decline

Indian region is endowed with wild diversity and genetic resources of many domesticated animals and avifauna. It is striking to note that biodiversity of only those breeds of domestic animals are reported which are specifically meant for consumption of their meat, eggs, milk, compost, dung the source of rural energy, wool, carrying loads and ploughing agricultural fields. Whereas the races and breeds of domesticated canines (dogs) and felines (cats) and pigeons have not been recorded under the genetic resources in the manner these species deserve. This appears due to the fact that by and large these groups of animals and birds are mainly and only used as favourite pets. Several of the breeds and strains of these group of animals are threatened as a result of genetic erosion, contamination, blood sanguinity, egg shell thinning due to residual pesticides in feed, invasion of exotic breeds etc. Although a large portion of population of such animals are non-descript and economically non viable but they carry with them the genes of potential traits for variety of purposes.

Domesticated Animals and their Wild Origin

It is interesting to note here that the domestication of animals came into being in the centres of ancient civilization. These centres are referred as 'primary centre' and thought to be in far east and central Asia. The success behind the outcome of a particular breed must have been due to collective efforts through the conscious/subconscious selection and forced admixture. During Neolithic - Chalcolithic period (*ca* 1800-1200 BC), it is revealed that in many parts of India cattle, sheep, goat, horse, ass, dog and pig were domesticated (Mehra, 1997). The main aim for animal husbandry, herding, hunting and fishing was to enhance food and animal production.

It may not be out of the way to mention here that the origin of several of the races and domesticated varieties of animal diversity existed in wilderness and still exists.

Humped Cattle: Following are certain facts about this group of domesticated animals.

- Domestication and rearing began during Harappan civilization (2200 to 1600 B.C.), Afghanistan, Sindh and Baluchistan of Pakistan.
- Fossils of *Bos acutifrons* was discovered in Siwaliks and considered to be the wild relatives of domesticated cattle.
- Zoogeographical distribution of the species from primary to secondary centre might have occurred through the interference of man.

Vechur Cattle

A miniature breed of Kerala is almost on the verge of extinction. The cattle is reputed to subsist on low and poor quality feed and yields 2 kg milk in a day during milching period (Sahai, 1999).

Buffalo

- Wild water buffalo (*Bubalus arni*) was extensively distributed in India in olden times. Swamp type buffaloes occur wild in NE India.
- The species was reared during Harappan period.

Yak

- *Paephagus mutus* is considered to be the wild species of origin of modern times race, for example, *Paephagus grunniens*. Both the species occur in common habitat in Ladakh and Tibet.

- Earliest domestication dates back to 1st millennium BC in Trans Himalayan region, China, Mongolia, Russia etc..

Mithun

- It resembles with Indian bison (Gaur) and is believed to be under domestication since the period of Mohanjo - Daro (2500 BC). Singh (1999) believed that the selection and domestication of the species depended on the quantity of meat for the consumption purpose. Due to free ranging, the selection norms for domestication of the species could not be as effective as in case of other animals.

Sheep

- *Ovis musimon* (Mouflons), *Ovis orientalis* and *Ovis vignei* are considered to be wild species of sheep found in the mountainous areas of Europe and Asia. It is believed that the domesticated sheep (*Ovis aries*) was brought to domestication nearly 9000 years ago. The various races differ by their size and body weight. While Indian domesticated sheep (*O. orientalis*) has wild relatives in the mountains of Afghanistan. *Ovis vignei* is distributed from Pakistan to Iraq. Taxonomic studies on the genus *Ovis* reveals that the 30-40 well identified races fall into Euroasian (*O. musimon,* and *O. vignei,* and *O. aegalis)* and American wild sheep (snow sheep, thin and big horns) groups.

Goat

- Domestication of goat dates back to 9th millennium B.C. It has over 10,000 years of history of association with mankind. Singh (1999) mentioned that the occurrence of European and Asian remnants of the species indicate that by Pleistocene times these animals had become much more common and closely related to certain species of *Capra* and *Hemetragus* still in existence. The domestic breed of the goat is the descendent of *Capra prisea* (Central Asiatic) and *Capra aegagurus* (Persian). The *falconeri* form of the latter is also considered to be the ancestor of Indian breed. The wild goats in general are divided into *ibex, typical* of *algagrus* and *falconeri* (markhor) forms.

Camel

- The taxon is represented by *Camelus bactriances* and *C. dromedaries.* Wild camels seem to have survived in the North America into Neolithic period. The domestication refers back to 3000 BC in Iran and was known to Indus civilization (2300-1800 BC). It is believed that the silk route from China through central Asia resulted in the spread of this species in the Indian region. A population of less than 100 individuals of double humped camel exotic to Leh region of Jammu and Kashmir is found in the country.

Pig

- Indian wild pig (*Sus cristatus*) is believed to be the ancestor of domesticated breeds. The former has distribution in the Terai, sub-Himalayan and Siwalik tracts of north Indian region.

Poultry

- Red jungle fowl (*Gallus gallus*), typical of different types of forest of the country is considered to be the originator of domesticated fowl. The history of domestication goes back to the Mohenjo Daro civilization.

- The poultry comprising various significant avian breeds such as chickens, ducks, guinea fowl, etc. are represented by specific habitats and ecosystem of occurrence typical of some regions of India as provided below:

- Northern India: Brown Desi (Uttar Pradeshand Uttaranchal), Faverolla (Jammu & Kashmir), Punjab Brown (Punjab and Haryana) and Teni.

- Northeastern Region: Daothigir (domesticated by Bodo community of Assam), frizzle Fowl (curly plumaged), Miri (by miri tribe), Naked neck and Teni.

- South India: Aseel (dogged fighting quality typical of Andhra Pradesh), Ghagus (domesticated by nomadic communities of Andhra Pradesh and Karnataka), Denki (Fighting quality, resistant to diseases - Andhra Pradesh), Tellichery (Meat with medicinal value - Kerala).

- Central India: Kadamkanath or Kalamasi (Internal morphology with purplish or blue black coloration - Madhya Pradesh), Teni.

- Nicobar islands: Nicobari.

Pigeons

- Domesticated breed of this pet bird is known to have evolved from wild *Columba livia* (Blue Rock Pigeon) which is found through out India upto 4200 m in the Himalayas. The earliest record of domestication of pigeon dates back to 3000 B.C. in the fifth Egyptian dynasty. It is mentioned in Charles Darwins', "*The Origin of Species*" (vide Evolution Website, Chapter I: Variation under Domestication, http://www.bbc.co.uk) that around the year 1600 Akbar, the Mogul Emperor kept 90,000 pigeons and developed breeds. a pigeon hobbyist (Fancier) maintains the genetic stock of breed. Phenotypically the bird is distinguished by the morphology of beak, wing span, eyes, legs, plumage, colour of the mentioned features etc. The specific life partnership of male and female pigeons has helped in keeping the different breeds together in the same aviary. According to Mr H. Biswas, a noted fancier of northern India, (*pers. Comm.*) the various prized breeds of the pigeon are commonly known as Kagzee (Spotless pure white with short black beak), Lakka (White albinos type with scarlet opaque eyes, crested and fantail), Kaldooa (white with black tail), phulsira (white with grey spotted tinges of plumage on upper half of the neck), etc. Mr. Biswas believes that some of the purest forms of breeds known for long distance flight (once involved in carrying secret message during world war times) are still reared in different parts of India. The breed with quality of flying into sky for hours together still remains a pride possession of many who earn the claims for the quality of the breed.

- The major threats to this group of domesticated bird are due to lack of awareness among the conservationists for inventorization and improvement of the breed. There is need to carry out studies on the reproductive behaviour, growing impact of polluted flyzone, egg shell thinning and removal of the practice of cross-breeding with impure forms.

Dogs (Pet)

- Indian dogs fall into nondescript, authentic breeds (registered) and cross breeds (unregistered). These dogs seem to have evolved from a genetic lot of as many as four groups such as *Canis familiaris intermedius, leineri, metris optimae*, and *mountain dogs* and *mastiffs*. According to Sylvester (1989) the aforementioned groups belong to primitive dogs represented today by the Asian Pariah and the Australian Dingo. In general the dog has its ancestary in *Canis lupus* (Wolf) with zoogeographical entity, origin and diversity with small wolves of Indian and Arabian regions. The large and small dogs typical of subarctic origin from large northern wolves gave rise to the breeds of large (*Akita, Alaskan Malanute, Siberian Husky* etc.) and small dogs (*Chow Chow, Finnish Spitz, Pomeranian* etc.). Some of the prized breeds of dog seen in Indian region are Cocker Spaniel, Poddle, Dobermann Pinscher, Greyhound, German Shepherd, Bichon Frise, Borzoi, Boxer Bulldog, Cocker Spaniel, Dachshund, Dalmatian, Keeshond, Japanese Chin, Labrador Retriever, Lhsa Apso, Tibetan Spaniel and Terrier etc.

Threats to the nondescript but carriers of most promising genes of several of the known breeds are arising due to neglect, non registration and development of stray nature. It deserves mention here that several phenotypes of run-stray of some good breeds are observed in various parts of the country.

Equine Breeds

- India is endowed with six distinct breeds of horses. Some of the following equines still exist in their wild habitats.

NW Region: *Kathiawari* and *Marwari*.

Trans Himalayan Region: *Spiti* and *Zainskari*.

Eastern Region: *Bhutia* and *Manipuri*.

FAO (see Pundir, 1999) reported additional Deccani, Chummarti and Sikang ponny from India.

While the population of mules has remained unchanged for the past over three decades there is pronounced decline in the population of horses and asses. It is largely due to better and rapid alternatives to means of transportation, social status, maintenance norms etc. The mules enjoyed the preferene of man for their use in tough terrains and climates.

Kathiawari Horses: Considered to have been originated from wild horses of Kathiawar. The breeding tract covers larger areas of Saurashtra region of Gujarat.

Marwari Horses: derived from Marwar region of Rajasthan covering Jodhpur, Udaipur etc.

Spiti Horses: Distributed in Kinnaur and Kullu districts and Spiti Valley of Himachal Pradesh. The breeds have the unique adaptation to low temperature, high altitude survival, scarcity of food etc.

Zainskari Horses: Typical of Zainskar and other Valleys of Jammu and Kashmir. Pundir (1999) while mentioning about the establishment of a Zainskari Horse Breeding Farm at Leh in Jammu & Kashmir reports that only a few hundred horses of this breed exist. This is largely due to a substantial breeding of the species with non descript animals.

Bhutia Horses: Distributed in Sikkim and Darjeeling Himalayas and has affinity with that of Tibetan race.

Manipuri: Typically distributed in Manipur and adjacent region of NE India.

Ass: Indian wild ass and kiang are found in Rann of Cutch in Gujarat and Sikkim & Ladakh respectively. The domesticated asses mainly descended from the Nubian race typical of NE African origin.

Threats to the Indigenous Domestic Animal Diversity

Besides the genetic erosion caused by influx of exotic breeds, there are several of the following reasons behind the depletion of the diversity and genetic resources.

1. The superior and high yielding milch buffaloes are taken to cities for the trade of milk. Once the lactation period is seized these animals are sold to slaughter houses. The characterization with reference to the specific and superior breeds is not carried out and the significant genetic diversity is lost into oblivion.

2. There is not only scarcity of superior breeds of males among the domestic animal diversity but also the site-cum-conservation specific approaches of the species. This covers the maintenance and further improvement of the pure breed.

3. Gene erosion of superior breed due to subconscious selection, cross breeding and admixture. The case is that of 'Bhadawari' buffalo of the erstwhile region of Agra, Etawa etc. crossed with 'Murrah' breed which subsequently resulted into the decline of the breed from 1000 to 30 and one in recent past (Sahai 2001 *pers. comm.*).

4. Due to shrinkage of the original area of grazing and place of gathering and the free movement of the superior bulls and other species the chances of breeding with indigenous superior breeds have reduced and become non existent.

5. This may also not be ruled out that the presence of pesticides, artificial fertilizer etc. in the residual forms of the feed and subsequently in the body result into low count of semen and egg shell thinning.

Acharya (1999) and Sahai (1999) mentioned about the decline of following indigenous breeds of livestock.

Livestock	Breeds
Camel	Double humped.
Cattle	Bachaur (Vechur), Danji, Kenkatha, Siri, Kherigarh.
Buffalo	Toda.
Sheep	Nilgiri, Mandya, Hissardale, Bhakarwal, Poonchi, Karnah, Gurej.
Goat	Barbari, Surti, Jamnapari.
Poultry	All indigenous breeds.

It is also feared that with the economic globalization and free trade under WTO (World Trade Organization) the chances of invasion of genetically manipulated breeds may dominate the indigenous elements in time to come.

Need and Priorities on the Biodiversity Conservation and Genetic Resource Management of Domesticated and Potential Animals.

1. Inventorization on the domesticated wild species of animals with reference to various indigenous communities (tribal).
2. Characterization of various phenotypes of non descript groups of species and profiling of various features of those species recently brought to domestication using recent trends.
3. Registration and development of computerized database of all the potential wild races of domesticated animal genetic resources with significant traits of interbreeding and economic importance.
4. Evaluation of rare and threatened categories of important species (breeds) with reference to site/specific norms. Identification of custodians of rare and pure breeds and development of promotional avenues for such keepers.
5. Development of inter-institutional networking and support with organizations involved in carrying out studies on some aspects of management, utilization and conservation of resources concerning the development of domestic animal biodiversity.

Literature Cited

Acharya, R.M. (1999): Management and breeding strategies for conservation and development of nation breeds: 21-27. In: R. Sahai "*Short Course on characterization and Conservation of Domesticated livestock and poultry resources*". National Bureau of Animal Genetic Resources (ICAR) Karnal - 132001. Haryana, India.

Mehra, KL (1997): Biodiversity and subsistence changes in India: The Neolithic and Chalcolithic age. Asian - Agri History 1(2):105-126.

Nivsarkar, A.E. (1999): Animal genetic resources of India vis-à-vis -past, present and future perspectives. *ibid.* 1-8.

Pundir, R.K. (1999): Equine genetic resources and their conservation in India.: *ibid.* 42.

Sahai, R. (1999): Conservation of domesticated animals - Conflicts & Contractions. *In*: R. Sahai "*Short Course on characterization and Conservation of Domesticated livestock and poultry resources*". National Bureau of Animal Genetic Resources (ICAR) Karnal - 132001. Haryana, India.: 28-33.

Singh, Gurmej (1999): Domestication in retrospect - process of speciation and emergence of breeds. *In* : R. Sahai "*Short Course on characterization and Conservation of Domesticated livestock and poultry resources*". National Bureau of Animal Genetic Resources (ICAR), Karnal - 132001. Haryana, India.: 9-16.

Sylvester, P. (Ed.) (1989): The Reader's Digest Illustrated Book of Dogs. Revised Edition pp. 384. The Reader's Digest Association, Inc. Pleasantville, New York.

Chapter IX

Marine Biological Diversity of India

The marine biological diversity of India falling in the Central Indian Ocean Marine Region covers Laccadives (Lakshadweep), Andaman and Nicobar archipelagos. Terrestrial ecosystem of Andaman and Nicobar contributes towards the major components of one of the 25 globally recognised biodiversity hotspots, that is, Indo-Myanmar. The countries adjoining to Indian marine region of biodiversity significance are Sri Lanka, Myanmar, Maldives, Bangladesh and Chagos Archipelago (British Indian Ocean Territory). The oceanic region is circumscribed by the part of the Arabian Sea, Bay of Bengal and area south of India. The Bay of Bengal has lower salinity due to high runoff and precipitation than that of the sites falling in the Arabian Sea.

Coral Reefs Biodiversity

The coral reefs of the region are known for all the three main types, namely atoll, fringing and barrier. The growth of reef is inhibited where there is massive fresh water and sedimentary inputs from different rivers. The Andamans and Nicobars have fringing and barrier reefs around many islands. Gulf of Mannar posses fringing reefs. The reefs of Laccadives are poorly known from biological diversity point of view (Wells *et al.*, 1995) with 9 genera in the northern atolls. Pande and colleagues (1991) reported the occurrence of coral diversity of 179 species distributed among 39 genera from Andamans and Nicobar out of 342 species covering 76 genera from whole of the country. Seagrass is found in Gulf of Mannar, Palk Bay, Laccadives and Andaman and Nicobars.

Species Diversity of Indian Coral Fauna

S.N.	Region	Genera	Species
1.	Laccadive	28	107
2.	Gulf of Kutch	25	38
3.	SE Coast of India	38	95
4.	West Coast of Kerala and Tamil Nadu	18	30
5.	Andamans & Nicobars	60	136

The Coral reefs have interdependent species of fauna with repre sentative of families *Chaetodontidae, Seridae, Labridae Pomacentridae* (damsel fish), *Serranidae, Holocentridae* an *Ostraciodontioidae*. The association of species contributes toward the productivity and diversity of coral reefs.

Seaweeds

Seaweeds comprising nearly 1000 species are members o Phaeophyceae or brown algae referred to as primitive grass of non flower ing plants (Thallophyta). These are commercially important living marin resources. The seaweeds grow submerged in intertidal, shallow and up t 100 m depth in the sea and brackish water estuaries. Commonly distribute along the Tamil Nadu and Gujarat coasts, Laccadive and Andamans & Nicobars. Out of 625 species of marine algae nearly 10% are comme cially important e.g. agar from red algae (*Gelichiella, Gracilaria*) an algin from brown algae (*Sargassum* and *Turbinaria*).

Planktons and other Marine Biological Diversity The phytoplankto of Indian ocean has Dinoflagellates with 452 species. Zooplankton is en dowed with 300 species in the Arabian sea and Indian ocean.

Seafan (Gorgonids): Seafan represents the phylum Coelenterate with 2 species recorded from Indian ocean. Export of seafans and seaferns show declining trend of 25 tons from 1975-79, 11 tons from 1980-84 and one to in 1990-91 and nil export in 1992-93 (Anon., 1993).

Prawns and Shrimps: They belong to different families of Natantia an comprise of 63 sps. of littoral and 27 species of deep sea forms, beside several occupying the inland water bodies. Dominant group among insho prawns are *Penaeus indicus P. penicillatus, P. marguiensis, P. monodo P. semiculcates, Metapenaeus dobsoni, M. monoceas, M. affinis*. Shrim (*Solenocera crassicornis*).

Kerala and Maharashtra have 3/4th of prawn population while Gujar Andhra Pradesh and Tamil Nadu posses 20%. Karnataka, Orissa, W Bengal share 5% while Goa, Pondicherry and Andamans 1%.

Lobsters - 17 species of lobsters inhabit coastal areas. 11 species are littoral and 6 species deep sea forms. Littoral species are represented by *Panulirus polyphagus* (Gujarat and Maharashtra), *P. homarus* and *P. ornatus* (Trivandrum, Kanyakumari).

Crabs: There are 640 species in Indian waters with 12 as edible, namely *Portumus sanguinolanto. P. pelagicus, Charybolis feriatus, C. lucifera, C. anmilates, C. nataton, S oylla t ranquebarica, S. serrata, Matuta lunarix, Sesarma tetragonum* and *Varuna littorates.*

Molluscs: India has a total of 3271 species belonging to as many as 220 families and 591 genera, out of which 1700 are gastropods, 1100 bivalves, 210 cephalopods, 41 ploypIacophs and 20 scaphopods.

Mussels: *Perna viridis,* green mussel occurs along Cochin, Malabar coast, Goa, Gulf of Kutch, and wizag to Andaman. Brown mussel *Perna indica* has limited distribution.

Clams: *Meratrix casta, Paphie malabarica, Villorita cyprinoides, Anadara granosa, Mesodesma glabral, Donax faba, Mactra violacea, Tridacana minima, T. crocea* are the important species. Kerala represents maximum clams in the region covering Vermbanad and Ashtamudi.

Bivalve Oysters: 8 species are reported from Indian waters of which 4 are commercially exploited e.g. *Crassostre madraensis, C. gryphoides, C. diocoides* and *Saccostrea cuculla, C. madraensis.*

The most important Indian edible oyster is distributed along Orissa (Sonapur) to Karnataka.

Window pane Oysters: *Placenta placenta* occurs in Gulf of Kutch, Mumbai, Malabar coast, Tutic Mandapam, Nagapatnam, Chennai and Kakinada Bay.

Pearl Oysters: There are 6 species of pearl oysters occurring along the Indian coasts. These are *Pinctada fucata, P. margarita, P. chemnitzi, P. anima, P. sugillata* and *P. annomoides.*

Natural Pearls have been collected from *P. fucata* which occurs in Gulf of Mannar, Tamil Nadu to Gulf of Kutch. *P. margenitifera* distributed in Andamans yields high quality pearls.

Giant Clams: Four species with edible property are known from Indian waters, namely *Tridacna crocea, T. maxima, T. squamosa* and *Hippopus hippopus* distributed in Andaman and Nicobar Island.

Gastropods: *Xancus pyrum* (Sacred Chank) is distributed in coastal waters of Tamil Nadu, Kerala, Gulf of Kutch and Andamans.

Trochus and Turbo: *Trochus niloticus* (Top shell) and *Turbo marmoratus* (Turban shell) occur in Andaman and Nicobar.

Cephalopods: Squids are commercially important cephalopods from Indian coasts. *Loligo duvancalli, Sapiotcultis lessoniana, Dorytenthis* sp., *Liliotus investigatoris.* cuttlejides are sepia *pharonis, S. aculeata, S. branmana, S. elliptica, S. prashadi, S. inermis.* **Octopus** - *Octopus dolyusi, O. oegina, O. membranaceus, Cystopus indicus* and *Berrya keralensis.*

Sea Cucumbers: There are 200 known species occurring in seas around India. Nearly 75 species are present in shallow waters while nearly 50 species in the inter-tidal region.

Sea cucumbers have distribution in the Gulf of Mannar from Mandapam to Tuticorin and in the Palk Bay from Rameshwar to Mallipatam , and some occur in Andaman and Nicobar Islands also.

Jelly Fishes: The jelly fishes are seen abundantly in the in-shore and backwaters of Tamil Nadu and Kerala during the months of January to June. Common among the species are that of *Crambionella stulhamani C. buitendijki, Dactylometra quinquicirsha* etc.

Sea Urchins: These species have niche on ocean floor, usually on rock bottom. Distributed in the Quilon - Kanyakumari region of the southwest coast of India. Species are *Salmacis bicolor, S. virgulata, Fibularia volva Echinostrephus molaris* etc.

Fishes: Bay of Bengal, Arabian Sea and Indian Ocean including coastal and islands have about 1440 species.

Holothuria scabra (Sand fish) and *H. spinifera* (Brown sand fish) once exploited from the Gulf of Mannar and Palk Bay are now threatened. Other species are *Actinopyga mauritiana, A. echinites, Bohadochia marmorata, B. argu, Holothuria nobilis, Stichopus chloronotus* etc.

Other Examples: Bombay Duck (*Harpodon nahareus*), Oil Sardine (*Sardinalla longiceps*), *Sardinalla fimbriatus*, Malabar sole (*Cynoglossus semnifasciatus*), sharks (*Cacuarshisms bleekeri*), Rays (*Narcine brunnea*), Silver bellies (*Laiognathus dussumieri*), Flying fish (*Exocoetus volitaro*), Eels. Tunas are present in large open waters. Ten of the marine fishes in India are categorized as threatened and referred as vulnerable and indeterminate, while 8 marine fishes are indeterminate.

Marine Turtles: Five species of marine turtles found in India are considered to be vulnerable to different biotic influences.

- Green turtle *Chelonia mydas* (Gulf of Kutch and Saurashtra, Peninsula, uninhabited islands in Laccadives).
- The hawksbill, *Eretmochelys umbricata* (Largest population in the Andamans particularly in South Reef and North Brother).
- The loggerhead *Caretta caretta* (Rare).
- The leather back, *Dermochelys coriacea* (Main populations in Andamans and Nicobar).

  Olive Ridley turtle (*Lepidochelys olivacea*) is most common. In Orissa the species has two important nesting beaches with 300,000-500,000 nests; other nesting sites are on the mainland coasts of Andamans and Nicobar, and Sunderbans.

Marine Birds: Several globally threatened marine and coastal species occur in India, Sri Lanka and Bangladesh. The sea birds (Seagull, Gannet, Albatross, Tern etc.) of this region are poorly known and do not appear to be abundant. Sunderbans are an important staging and wintering area for gulls and tern. Pitti and Baliapani in the Laccadives may have sea bird colonies.

Mammals: There are 22 catalogued species of marine mammals frequenting the coastal waters mostly for feeding or breeding around Indian seas.

Whales: Scientific data on whales from India is scanty and the available knowledge is based on information from a few dead carcasses on their entangling in nets.

(i) Toothed Whales: Sperm whale, *Physeter macrocephalus*; Pigmy sperm whale, *Kogia breviceps*; Cuvier's Beaked whale, *Ziphis camostris.*

(ii) Baleen whales: Blue whale, *Balaenoptera musculus*; Sei whale, *Balaenoptera borealis*; Fir whale, *Balaenoptera physalus*; Humpback whale, *Megaptera noveangliae*; Brydes whale, *Balaenoptera edeni*; Mink whale, *Balaenoptera acutorostrata*, Australian whale, *Balaena australis*

Dolphin: *Delphinus delphis* is reported from coastal waters of Andhra Pradesh, Tamil Nadu and Gujrat. Ocean population of humback dolphin (*Sousa chinensis*) and the spotted dolphin, *Stenella attenuata*, distributed along the mangroves are considered at risk.

Dugong: Sea cow (*Dugong dugong*) is reported from western coastal area, the Gulf of Mannar, Palk Bay around Andaman islands (Ritchie's Archipelago and North Reef).

Species Diversity of Marine Ecosystem (Indian Ocean)

Taxonomic group	Family	Genera	Species	Status
Phytoplankton			452	
Zooplankton	2		322	
Algae			624	
Sea grasses			11	
Higher Plant				
Mangrove			45	
Invertebrates				
Sponges			305	
Coelentenates			214	

Crustaceans			764	
Molluscs	220	591	3271	
Echinoderms			200	
Vertebrates				
Fish			1440	2 vulnerable & indeterminate
Reptile (Turtles)			5	All at risk
Mammals			22	From endangered to rare
Total diversity	**222**	**591**	**7675**	

Marine Biological Diversity of Gulf of Mannar at A Glance

Biodiversity	Groups	No. of Species
Flora	**Algae**	150 (624)
	Red Algae	
	- *Gelidiella*	
	- *Gelidium*	
	- *Gracilaria*	
	Brown Agae	
	- *Sargassum*	
	- *Turbinaria*	
	- *Hormophysa*	
	- *Macrocystis*	
	- Ascophyllum	
	Green Algae	
	- *Ulva*	
	- *Enteromorpha*	
	- *Caulerpa*	
	- *Codium*	
	Seagrasses	10
	- *Halophila*	
	- *Cymodacea*	
	- *Halodule*	
	- *Thalassia*	
	- *Syringodium*	
Fauna	Sponges	280
	Corals	92
	Annelids	40(27 g 14 f)

Biodiversity	Groups	No. of Species
	Shrimps	24 (5g)
	Sea fans	22
	Polychaetes	160
	Prawns	35
	Brachyuran craks	106 (57 g/17f)
	Lobsters	
	Gastropods	119 to 178 (76 g/45f)
	Bivalves	75 (47g/28f) to 156
	Cephalopods	17
	Echinoderms	103
	Fishes	90 (26f/4 orders)
	Coral fishes	8
	Turtle	2
	- Green Turtle	
	(Chelone mydas)	
	- Hawksbill turtle	
	(Erethinocheles imbricata)	
	Cetacea	2
	- Toothed whale	
	- Baleen whale	
	Bacteria	
	- Heterotrophic bacteria	35
	- Heterotrophic ammonifying	22
	- Nitrate reducing	16
	- Free living N2-fixing	2
	- Phosphate solubilizing	24
	- Phosphate - producing	26
	- Arthrobacter	5
	- *Azatobacter*	2
	- *Bacillus*	11
	- *Breevibacterum*	1
	- *Flavobacterum*	1
	- *Flavomonas*	1
	- *Kurthia*	1
	- *Micrococcus*	4
	- *Pseudomonas*	5
	- *Sporosarina*	1
	- *Vibrio*	3
	Phytoplankton	126

Biodiversity	Groups	No. of Species
	Diatom	
	- *Coscinodiscus*	8
	- *Pleurosigma*	8
	- *Navicula*	7
	- *Triceratium*	5
	- *Bacteriastrum*	4
	- *Amphora*	3
	- *Fragilania*	3
	- *Gyrosigma*	3
	- *Nitzschia*	3
	- *Synedra*	3
	- *Thalassio Thrix*	3
	- *B idaluphia*	2
	- *C yclotalla*	2
	- *Melosira*	2
	- *Odontella*	2
	- *S keletonema*	2
	- *Thalassiosera*	2
	- *Thalassionema*	2

Based on Kannan *et al.* (1998); Kumaraguru and Jayakumar (1998); Subramanian (1998)
g = genera; f = families

Threats to the Marine Biological Diversity

- Poaching for commercial purpose
- Para - traditional professional activities (harvesting the genetically rich wild fauna)
- Sea weed harvesting through unsustainable means (Removal along with holdfast from the sand/germinating sprouts etc.)
- Commercial fishing of fauna
    - # Sea horses (*Hippocampus kuda*) for prostoglandin
    - # Sea fans and sea ferns
    - # Sea cucumbers as food

Soft corals

- Use of dynamites for fishing
- Removal of mangroves for aquaculture development, fuel, tannins etc
- Use of metal scrapers and other implements damaging the niche
- Fly ash discharge from Thermal Power Stations (Tuticorin - destroy pearl oyster fishing through the choking of gills and alimentary canals causing large scale mortality of the recruitment stock.
- Effluent discharge from chemical industries
- Direct sewage disposal, discharges from shipping, industrial waste, sediment nutrient run off.
- Reef rock mining
- Development of artificial ecosystem by aquaculture farms
- Degree of exposure and turbidity of waters due to coral mining and in flow of effluents.
- Uncontrolled tourism development
- Sea level rise due to global warming
- Soft bottomed habitats of seas are impacted by harbour developments which result into reduced water circulation and the resultant concentration of pollutants.

International Conventions for the Conservation and Establishment of Marine Protected Areas (MPAS)

The Law of the Sea Convention of 1982, deals with the conservation and management of the living resources. Under Article 117 of the Convention urges the countries to cooperate in taking such measures for their respective nationals, as may be necessary for the conservation of the living resources. There is need to take as per the measures adopted by UN Convention on the law of the Sea (UNCLS) best scientific evidence available to the state concerned to maintain and restore populations of harvested species at levels which can produce the maximum sustainable yield, as qualified by relevant environmental and economic facts including the special requirements of developing nations and taking into account fishing patterns, the interdependence of stocks and are generally recommended international minimum standards whether subregional, regional or global.

The ASEAN Agreement on the Conservation of Nature and Natural Resources for South East Asia, 1985 adopted a very comprehensive approach to the problem of preserving biological diversity. The parties are urged to establish the protected areas or the enactment of legislation to prohibit the taking of endangered species. The ASEAN Agreement requires parties to develop national conservation strategies as a programme to attain the goal of sustainable development. This pertains to great extent the cooperation for the establishment of marine protected areas particularly in marine areas beyond the limits of national jurisdiction.

Other conventions concerned are Convention on Biology Diversity (CBD) addresses issue on Marine and Coastal Biological Diversity. Protected areas (PAs) on the subject is one of the thematic areas and dealt as a cross cutting issue of CBD (Article 8 a to e) Convention on International Trade in Endangered Species of Wild Fauna and Flora (CITES), Convention on the Conservation of Migratory Species of Wild Animals, Bonn (1979), Antarctic Treaty of 1959, the Convention for the Conservation of Antarctic Seals of 1972 and the Conservation of Polar bears of 1973. The Conference on the Conservation of Antarctic Marine Living Resources of 1980 (CCAMLR) was adopted for the conservation of biological diversity at the regional level. The Contracting Parties recognized the importance of safeguarding the environment and protecting the integrity of the ecosystem of the seas surrounding Antarctica; which is full of marine resources. The United Nations Convention on Environmental Impact Assessment in the Transboundary Context, 1991, and several resolutions of General Assembly of the United Nations were adopted time to time with a view to protecting environment and preventing the contamination of the environment from the several pollutants which are injurious to human health and the environment including biological diversity. Different conventions are indirectly helpful in the conservation and protection of biological diversity which are adopted with a view to protecting general human biosphere in general and marine protected areas in particular.

Marine Protected Areas (MPAS) of India and their Jurisdiction in Adjacent Countries

With regards to the options for cooperation for the establishment of

MPAS beyond the limits of national jurisdiction following biogeogaphic zonation of the Central Indian Ocean Marine region is required to be studied: Western Indian Ocean (4 No.), Eastern Indian Ocean (8), Northern Bay of Bengal (2), East Bay of Bengal (1) while the Central Indian Ocean has none. Several of the species find congenial ecological corridors under water and through land. Among the 8 MPAs in Eastern Indian Ocean 4 are located around Sri Lanka and 4 along east coast of India. Wandur Marine NP located in the Andaman Islands is in the East Bay of Bengal. Sundarbans NP and Bhitarkanika WS are located in the northern Bay of Bengal Zone. From the biological diversity of transboundary consideration Sundarbans is a vast complex of intertidal and estuarine areas situated on the border of India and Bangladesh where the Ganga, Brahmaputra and Meghna rivers converge in the Bengal Basin. About 65 percent of the Sundarbans lies in Bangladesh and 35 percent in India. The area includes the most extensive areas of mangrove present in India (constituting more than 60 percent of the area of mangroves). India has nearly 3,900,000 ha of estuarine wetlands (134 brackish and 19 coastal wet land areas). The country has 1593 Marine and Littoral protected Areas (http://marine.wri.org/pubs.The atoll groups of the Laccadive - Chagos Chain and the Myanmar coast are yet to be represented.

Coral reefs have global and regional importance. These mainly pertain to India, Maldives, the Chagos Archipelago (British Territory), Laccadives and the coastal wetlands of India, Bangladesh and Myanmar. The Gulf of Mannar Marine National Park was established in 1980 to include 21 islands and the surrounding five kilometers of sea adjoining the region of Sri Lanka. Gulf of Kutch Marine Sanctuary and Marine National park of India from a large inlet of the Arabian Sea, covering approximately 7,350 square kilometers and with a maximum depth of about 60 meters. Most of the Gulf is criss-crossed with mangrove-lined creeks, the intertidal zone being sandy or muddy, or of exposed limestone. Numerous islands are present and most of these support fringing reefs, although these are of low diversity with 44

Marine Protected Areas (National Parks and Sanctuaries)

Category I (Fully in marine environment, mangroves, corals, estuaries, algal beds etc.) (Based on L. Kannan, *personal communication,* 2005)

S.N.	Protected Areas	Area (ha)
1	Mahatma Gandhi NP, Wandoor, South Andaman	28,150
2	Rani Jhansi Marine NP (Richis Archipelogo) Andaman	25,614
3	Lohabarrack (Salt water Crocodile) Sanctuary, south andaman	10,000
4	Gulf of Kachchh Marine NP, Jamnagar, Gujarat	16,289
	Marine Sanctuary, Gulf of Kachchh, Jamnagar, Gujarat	29,503
5	Malvan Marine Sanctuary Sindhudurg, Maharastra	2,912
6	Bhitarkanika NP, Cuttak, Orissa	14,500
	Bhitarkanika Sanctuary, Kendrapara, Orissa	67,200
7	Gahirmatha Marine Sanctuary, Kendrapara, Orissa	1,43,500
8	Chilka (Nalabund) WLS, Khundra, Puri, Ganjam, Orissa	1,550
9	Gulf of Mannar Marine NP, (Ramanathapuram/tuticorin) Tamil Nadu	623
10	Pulicat Lake (Bird) Sanctuary, Tiruvellore, Tamil Nadu	15,367
11	Point Calimere Sanctuary, Nagapattinam	1,726
12	Coringa Wildlife Sanctuary, East Godhavary, Andhra Pradesh	23,570
13	Krishna Wildlife Sanctuary, Krishna/Guntur, Andhra Pradesh	19,481
14	Pulicat LakeBird Sanctuary, Nellore, Andhra Pradesh	50,000
15	Sundarbans National Park (Tiger), North & South 24-Pargana, West Bengal	133,010
16	Halliday Sanctuary, (South 24-Pargana), West Bengal	595
17	Lotaian Sanctuary, (South 24-Pargana), West Bengal	3,800
18	Sajnakhali Sanctuary, (South 24-Pargana), West Bengal	3,624
	Total	623,630

Category II (Major part in sea) (Based on L. Kannan, *personal communication,* 2005)

S.N.	Protected Areas	Area (ha)
1.	North Buttan NP., Middle Andaman	44
2.	Middle Buttan, Middle Andaman	44
3.	South Buttan N.P., Middle Andaman	03
4.	North Reef Island Sanctuary, North Andaman	348
5.	South reef Island Sanctuary, Middle Andaman	117
6.	Cuthbert Bay Sanctuary, Middle Andaman	582
7.	Cingue Sanctuary, South Andaman	951
8.	Galathea Bay Sanctuary, Great Nicobar (Larger part of Galathea National Park is in Terrestrial Ecosystem)	1,144
9.	Parkinson Island Sanctuary, Middle Andaman	034
10.	Mangrove island Sanctuary, Middle Andaman	039
11.	Blister Island Sanctuary, North Andaman	026
12.	Sandy Island Sanctuary, South Andaman	158
13.	Pitti wildlife Sanctuary, Lakshadweep (Laccadives)	01
	Total 3,491	
	Grand Total-33 National Parks and Sanctuaries (31 MPAs)	627,121

Needs and Priorities (Local, Regional and Global levels)

- Coast guard to be legal Guardian of Core Zone
- Aerial survey to locate coral reef patches and to enumerate marine mammals (Dolphins, whales and dugongs) through synaptic views.
- Implementation of eco-friendly aquaculture.
- Eco-friendly disposal of flyash
- Sand quarrying (Vembar and Tuticorin)
- Need for community - based resource management plans
- Conversion of conventional identification keys into figurative forms (diagnostic characters given in the form of figures side by side).

- Conversion of conventional identification keys into figurative forms (diagnostic characters given in the form of figures side by side).
- There is need for support and cooperation for the management of MPAs for providing breeding sites to different species such as Turtle (Green Olive Ridley, Hawksbill, Leatherback, Loggerhead) found in the sea of India adjoining Bangladesh, Sri Lanka, Maldives and Chagos Archipelago.
- Need for detailed analysis to determine whether the existing MPAs with disjunct elements of phyto/zoo geographical relevance significantly conserve a representative range of the habitat and ecosystem types.
- Environmental impact assessment with reference to the loss of biodiversity and genetic resource of common interest.
- Need for improvement in the management of existing MPAs particularly in the southern part of India from different threats.
- Countries among the SAARC, ASEAN, etc. are to develop inter-regional, inter-organizational, interdisciplinary research support and cooperation through ecosystem approach and specific species conservation programmes.
- Priorities species for conservation and management are turtles, dugong, fish, cetacean whales, humpback dolphin (*Sousa chinensis*), spotted dolphin (*Stenella attenuata*) migratory birds (Spot billed Pelican, Lesser Adjutant etc.).

Literature Cited

Pande, P.; A. Kothari and S. Singh (Eds.) (1991). Directory of National Parks and Sanctuaries in Andaman and Nicobar Islands. Indian Institute of Public Administration, New Delhi.

Wells, S.; S. Singh and R. Ivan (1995). Marine region 10: Central Indian Ocean - A report to the World Bank Environment Department. A Global Representative System of Marine Protected Areas. 16-38, Great Barrier Reef Marine Park Authority, The World Bank. The World Conservation Union (IUCN).

Anonymous (1993). Statistics of marine products exports. The Marine Products to Export Development Authority: 1-116.

Kannan, L; K. Kathiresan and A. Purushothaman (1998). Biodiversity of microbial flora (Bacteria and Phytoplankton) of the coral reef environment of the Gulf of

Mannar. In: Maikhuri, RK; KS Rao and RK Rai (Eds.) Biosphere Reserves and Management in India. Himavikas Occasional Publication No. 12: 207-220. GB Pant Institute of Himalaya Environment and Development, Almora, India.

Kumaraguru, AK and K. Jayakumar (1998). Under water ecology of coral reefs and their associated fauna in the Gulg of Mannar. In: RK Maikhuri; KS Rao and RK Rai (Eds.) Biosphere Reserves and Management in India. Himavikas Occasional Publication No. 12: 221-226. GB Pant Institute of Himalaya Environment and Development, Almora, India.

Subramanian, P. (1998). Secondary productivity and larval resources of Gulg of Mannar Biosphere Reserve. In : RK Maikhuri; KS Rao and RK Rai (Eds.) Biosphere Reserves and Management in India. Himavikas Occasional Publication No. 12: 227-236. GB Pant Institute of Himalaya Environment and Development, Almora, India.

Chapter X

Integration of Traditional Knowledge Resource and Grass Root Innovations for Biodiversity Registration and Conservation

Inventorization, characterization and subsequent registration of components of biodiversity involve the methodology of measurement, taxonomic evaluation, analysis and understanding. The registration goes through the process of data collection, analysis, authentication and conformation with authentic voucher specimens as represented in herbarium, museum record, field germplasm of natural or artificial areas of occurrence/ introduction. Various steps leading to effective registration need galvanizing with the recent trends of collection, recording, storage, retrieval and dissemination. The criteria and indicators in this regard need the involvement of experienced descriptor.

India occupies 2.4 per cent of the total geographic area of the world but supports about 8 per cent and 7 per cent of the total global flora and fauna respectively. Therefore, it ranks as one of the twelve-mega diversity centers of the world. The country has 89 National Parks and 489 Wildlife Sanctuaries representing about 4.70 per cent of total geographical area. There are 306 preservation plots (184 in conservation and 122 in production forests) with certain groups of communities inhabiting around. The complexity of diversity and phenotypically or genotypically significant stands of crop occurring in such gene pool present the survivors of various influencing factors such as climate change, global warming, pollution, imbalances in the equilibrium of interdependent dynamics, change in soil fertility and subsequent nutrient cycling etc. The forest communities dwelling nearby such key areas of conservation (KAC) have specific knowledge on various aspects of plots and their components that can be harnessed through benefit sharing. KAC to a larger extent represent certain indigenous communities who have associated with the activities of biodiversity conservation (BC) for several generations.

India is endowed with a sizeable geographic area under the forest cover, which provides us various resources and maintains the life support system of the nation. Therefore, if the forests are managed and maintained properly, a major lot of problems will be within manageable limits. In general it has become obligatory on the part of the global "*think tank*" to devise and adopt suitable ways and means to strike almost normal situation between the man and nature.

People's Participation and Conservation Strategies

Joint Forest Management (JFM) being an innovative approach has shown considerable potential and result to redress our environmental degradation from the impacts of the anthropogenic and biotic pressure on the rich biodiversity of the country. JFM is aimed at conserving our forest wealth with the participation of local people in a socio-political regime involving an amalgamation of their indigenous traditional knowledge and scientific modern forestry approaches; besides carrying out species level inventories of different phytogeographic and agro/eco/ bio-climatic regions of the country.

A considerable portion of our population lives in and around forest areas and appreciable proportions of the forest produce is consumed by them in different ways viz. food, fodder, dyes, fibre, gums, resins, medicines, etc. to carry out their living without much interaction with outside communities. Ethnic groups or the indigenous communities have accumulated a vast knowledge through their traditional ways of survey, collection, identification and usage of plant species. This treasure of information passes as oral scripture from one generation to another and they make all possible efforts to pass it on to the civilized world and getting documented and publicized.

Indigenous Traditional Knowledge (ITK) and Resource Management

India has nearly 580, 000 number of villages out of which 170, 379 encompass 32, 198, 305 hectare area of forest (FSI Report 1999). These villages representing over 54 million ethnic people in 430 distinct ethnic and over 227 linguistic groups. It indicates that the dependence of people on the forest biodiversity and genetic resources for years together might have developed and enhanced the traditional knowledge on various aspects of conservation. The traditional knowledge on the biodiversity of the country is as old as Neolithic and Chaleolithic age (2600 to 800 BC). It is evident om the discovery of remnants, relicts and wood charcoals (Mehra, 1997). ne cultures of that period was based on cultivation of wheat (dwarf - *iticum sphaerococcum;* club - *T. compactum;* bread - *T. aestivum;* nmer - *T. dicoccum,* barley, lentil, field pea (*Pisum arvense*), sorghum *orghum bicolor*), wild rice (*Oryza rufipogon*), pearl millet (*Coix cryma - jobi*), sesame, linseed and safflower (*Carthamus tinctorius*) c. Wood charcoals from the Pre-Harappan and Harappan periods belonged *Capparis, Cedrus, Toona, Manilkara, Acacia, Shisham, Tamarix, ziphus, Prosopis specigera, Lawsonia, Tectona, Calligonum, Ficus xburghii,* Peepal, *Terminalia, Albizia, Anogeissus, Morus* etc. There e evidences of the subsistence economy based on hunting and animal sbandry. Indigenous traditional knowledge exists on 10000 wild plant ecies including 7500 species for medicinal purpose, 3900 subsidiary food, 5 fibre and cordage, 400 as fodder, 300 piscides and pesticides. hnobiological studies document on 76 species of fauna (60 vertebrates - nphibians, birds, pisces, eptiles and mammals; 16 invertebrates - arach, ls, insects, molluscs etc. AICPE (1999). Indigenous traditional knowledge ists on 10,000 wild plant species including 7500 species for medicinal rpose, 3900 subsidiary food, 525 fibre and cordage, 400 as fodder, and 0 piscides and pesticides. Ethnobiological information exists on 76 species fauna (60 vertebrates - amphibians, birds, pisces, reptiles and mammals; invertebrates - arachnids, insects, molluscs etc.AICPE (1999). It would t be an exaggeration to say that the indigenous traditional knowledge K) of different communities, if could be shared and integrated with vances on the subject of biodiversity conservation over a period of time, stainability and management of biodiversity and genetic resources would much beyond expectations. In spheres where indigenous knowledge of al people is involved in manufacturing some value added process and duct they should be made share holders in profit. This exercise, besides viding them some monetary gains will infuse a sense of responsibility l ownership among them and will, therefore, form a keystone for the ured success of the biodiversity and genetic resources management.

The significance of mutual benefits through the integrated approach sharing of wisdom of indigenous knowledge (IK) and advance knowledge K) holders, that is, IK+ AK= IA (strengthening of participatory will) + (Integration of knowledge for application) may be deduced so. IA largely ends upon motivation, awareness, involvement, sharing of benefits and ponsibilities and 2 K needs specific approach and integration.

Indigenous Traditional Knowledge and Disaster Management

Indigenous communities particularly the ethnic communities hav knowledge on the flora and fauna which they do use in crisis such as durin lean season of food scarcity, draught, flood; construction of shelters at th time of environmental disasters. Knowledge on the behaviour of speci and the surroundings helps in the management of disaster in advance. F example, in the recent tsunami disaster (26th December 2004) whic affected the coastal region of Andaman and Nicobar tremendously, th ethnic communities sensed the crisis in advance through their ways interpreting the behaviour of fish, sea gulls and insects. They moved inlan relatively on higher grounds and made shelter on the tall poles of trees avoid the tsunami waves.

Biodiversity Registration

For the lack of documentation and registration of traditional knowled as grassroot innovations/inventions/discoveries the matters concerni intellectual property rights/patent and legal issues are underscored a devoid of due benefits to the knowledge holders. The features deservi authentication and subsequent registration are needed to be covered und following step wise heads.

* Inventorization
* Involvement of Indigenous Communities (IC)
* Authentication of knowledge
* Interpretation
* Integration of ITK with AK
* Innovation/Invention/Discovery
* IPR registration
* Incorporation
* Accessibility/publicity
* Implementation

Key issues on participation of IK holders for biodiversity innovati through integration of Advance Knowledge (AK) are mainly based up following categorization and mode of participation:

Identification of Components

- Biodiversity Conservation Significance
- IK holding communities
- Pros and cons (problems)
- Implementing agencies (Action)
- Collaborating Agencies.
- Village of indigenous traits

Priority areas with respect to Indigenous Communities

* Watersheds
* Physiographically/Politically demarcated areas
* Key areas of conservation

Source of potential information on the use of IK on biodiversity and participation of communities for the development of products (Coded) and underdeveloped/under-utilized/under explored components (Non coded) for biodiversity conservation would largely depend upon the rapid retrieval and dissemination of non-coded information base.

Coded

- Indigenous knowledge of ethnic groups coded. (scientifically studied upto efficacy/use level.)
- Traditional groups
    * Foresters
    * Scientific community
    * Medium
- Rural (Farmers) of the area/of the area placed elsewhere.
- Urban or Suburban
- Medicinal uses û Ayurveda

- Siddha
- Naturopathy
- Vaidyas (Kabiraj)

Local medicines

Modified/Improved Synthetically

Non Coded

- Ethnic (under explored/unexplored of region of group)
- Communities from other areas.
- Scientific groups
- * Scientists/biologists/conservationists
- * Socio-economists
- * Newly settled from other areas/scientific group - scientists, conservationists, biologists and others).
- * Traders without recourse to the identification of products and sources.

Further, indigenous knowledge on forest species of environmentally sensitive areas has been least converted into scientific language and brought to the application and integration. The identification, functioning, usage of several of such components for the conservation point of view are well known to the indigenous communities dependent on them. Indigenous knowledge about tending the river, sustainable utilization of land resource, preservation of flora and fauna, development and maintenance of ecological important sites such as sacred grove, preservation plots and other nature reserves, etc. would find most important entry into the Biodiversity Register to be made at local, regional and national levels.

Some Examples for Interpretation of ITK for Integration and Incorporation in envisaged Register and Documentation.

Hunting - Faunistic diversity
- Ground truthing (survey)

PLATE IV

ETHNIC DIVERSITY OF INDIA

1. An Andamese tribal hunting fish with bow and arrow
2. Tattooing with herbal dyes
3. Khasi tribals in festive mood
4. Andamanese tribals yachting in sea

- Species (selection, population, criteria)
- Season (Lean, Need)
- Conservation aspects
- Uses and Usage (Meet, bone, Ivory, horn, fur, captivity/rearing, breeding etc.)
- Use of materials
- Group/individual (operation)
- Species

Gathering (NTFP + Food).

- Species diversity.
- Selection process (conscious/subconscious).
- Localities (CGD, KAC, Bioclimatic).
- Time (Phenological/Lean Season).
- Operative approach (individually/in group).
- Maintenance of viability, dormancy, etc. of germplasm.
- Tendency of cultivating/multiplying/conserving depleting resource.

Fishing

- Communities involved.
- Aquatic ecosystem/watershed.
- Choice of species (sustainable approach).
- Traits applied.
- Materials (forest based) used.
- Time of fishing and sustainability.

Pastorization

- Inventorization of landscapes with pastures lands and migratory routes.
- Characterization of species.
- Livestock/animals grazing

- Species of pasture land
- History of pastureland with potential/species (to be introduced)
- Aspects of conversion of nomadic attitude to settling
- Educational aspect.

Shifting Cultivation (Slash and Burn Agriculture/Jhum/Podu/Dabiya/Chasa)

- Soil-cum-forest fertility for conversion of existing stock to desired type.
- Increasing/reducing the jhum cycle.
- Choice of species being grown.
- Indigenous elements coming up after the jhummed areas are abandoned (Eco-morphological features).
- Degree of subsistence and attitude for permanent agriculture with introduction of forest species of economic and conservation value.

Some examples of Outstanding Traditional Knowledge for Documentation

Arunachal Pradesh

Mishmis (Bebejia)

- Man and Women
- Agriculture
- Trade (deer musk, wild medicinal herbs - *Coptis teeta*)

Wanchos (Wangsus and Wangsas) of Tirap district

- Village council (disciplinarian attitude towards BC)

	-	Weavers (art is restricted to the chiefs, families)
	-	Wood carvings
	-	Tattooing (herbal based)
Apatanis	-	Artistic craftsmanship
Monpas	-	Weaving, dyeing, paper making, painting carpentry, sculpture, pottery and bamboo works, shifting and permanent type of agriculture, Animal husbandry (Yaks, Mithun, Cows, Sheep, Pigs, Mules, Fowls).
Bihar and Jharkhand	-	Groups - Kolarian and Dravidian.
	-	Sacred grove (Jher - Worshiped) - Cluster of trees
	-	Making of mats, ropes.
	-	Wild fruits collection*
	-	Collection of lac*, rearing of silk worms* (*Provide potential for Biodiversity Conservation).

Uttar Pradesh and **Uttaranchal**

Tharus	-	Intimate knowledge of cultivation in Terai land (tillers of the soil)
The Korwas	-	Primitive beliefs and habits
	-	Subsist on forest produce.
	-	Indulge in shifting cultivation.

The Sahariyas (Banarawats)

	-	Wood cutting
	-	Gathering of forest produce
	-	Conversion the shifting cultivation (Dahaiya) into permanent agriculture.

Application of knowledge derived from the traditional and conventiona methodologies needs integration of advances for all incorporation an registration purposes. Certain plant groups such as bamboo, rattans an palms, and species of food and nutrition, medicinal value, etc. receive greater attention of advances in the field of science and technology durin the last decade particularly in the field of inventorization, characterizatior value-added products, propagation, marketing etc. and these are require to be integrated and worked out for the best line of action.

IK (indigenous knowledge) holders have often been exploited fc their wisdom on the economically important and environmentally significar components of nature. For a newer paradigm, there is urgent need to involv the IK holders for the dissemination of their outstanding traditional knowledg through the development and the use of national level registeration.

Need and Priorities (Key Issues)

1. Identification of IK holders and traits for AK integration.
2. Characterization of specific areas of IKH significance.
3. Mode of participation. It should be based on IK cum Biodiversit Conservation significance.
4. Development of database on IKH cum AK integration.
6. Involvement of workers (IKH) for joint venture on biodiversity base innovations considering the specific traits they possess.

For example:

- Weavers and wood carvers for craftsmanship.
- Mining and smelting for indicators of reserve in forest.
- Silkworm rearing and lac collection for the developmer of target species.
- Tattooing for herbal dyes and colors.
- Piscidal plant products for biopesticide properties.
- Bamboo and rattan works for development of species. (Fc certain areas tribal culture = bamboo culture).
- Sacred grove for conservation of gene pool.

7. Inter-disciplinary, inter institutional networking and support at local, regional, national and international levels of cooperation for the development of national level registeration.

Comprehensive Data Sheet on Forest Biodiversity

[For incorporation of findings on the forest biodiversity in Biodiversity Register based on grass root innovations and knowledge]

1. SYSTEMATICS

Entities of elements after the removal of Nomenclatural and taxonomic ambiguity:

Scientific name
Basionym
Synonym (Important)
English name
Vernacular name

(In areas of historical or recent range; besides the taxon with one name but different description and taxon with different available names in use but one described)

Trade name, Salient features (Diversity, taxonomic, phenology conservation value)

2. HABITAT

Forest type
Associates (upper, lower, undergrowth)
Area of occupancy

3. DISTRIBUTIONAL RANGE (As per IT)

General (Global)
India (National)
Region (State)
Local

4. POPULATION

Continuous
Disjunct
Endemic
Relict
Extent of occurrence
Area of occupancy
Individuals

5. USEFULNESS

(Known & potential)

Entire
Foliage
Flowers
Fruits
Roots

6. STATUS (IUCN categories based on site and species specific population and distribution criteria)

1. Extinct (Ex)
2. Extinct in the wild (EW)
3. Critically endangered (CR)
4. Endangered (EN)
5. Vulnerable (VU)
6. Threatened (T)
7. Rare (R)
8. Low Risk (LR)
9. Conservation Dependent (CD)
10. Indeterminate(Inadequacyof knowledge) (I)
11. Data Deficient (DD)
12. Not Evaluated (NE)
13. Least Concern (LC)
14. Reproductive constraints (REP)
15. Out of Danger (O)

8. DATABASE

- Field Survey
- Forest Flora

- Monographs
- Herbaria
- Ethnobotanical knowledge
- Database
- Artificial intelligence (biologists, botanists, foresters and conservationists)
- Knowledge Management
- Photographs
- Maps(Aerospace/remotely sensed/GIS)

9. REMARKS

10. Important References

Literature Cited

AICPE (1999): Ethnobiology in India - a status report - AICPE (All India Co-ordinated Project on Ethnobiology). Ministry of Environment and Forests, Govt. of India, New Delhi.

Biswas, Sas.; S. Chandra and SS Jain and M. Pal (2005): Indigenous knowledge on Himalayan flora for conservation and benefit sharing. *In:* Rawat, JK; S. Srivastava; Sas. Biswas and HB Vasistha (Eds.), "*Conservation of Biodiversity in India- Status, Challenges and Efforts*": 152-156, ICFRE, Dehradun, India

Biswas, Sas., Sumer Chandra and S.S. Jain (2000) Widening the horizon of joint forest management (JFM) through participation of indigenous knowledge (IK) holders on Indian forest biodiversity conservation (IFBC) presented in Joint Forest Management International Workshop, "*Decade of JFM - Retrospection and Introspection*" organized by Ministry of Environment and Forests and ICFRE, Govt. of India, 19th-20th June 2000 at Vigyan Bhawan, New Delhi.

Mehra, KL (1997). Biodiversity and subsistence changes in India: The Neolithic and Chaleolithic Age. *Asian Agri - History* 1(2):105-126. Asian Agri-History Foundation.

Biodiversity Conservation with Reference to Impact Assessments

Convention on Biological Diversity (CBD) in Article 14 provides following text relevant to the impact assessment with a view to minimizing adverse impacts. The part of the text is as following:

Article 14: Impact Assessment and Minimizing Adverse Impacts

1. *Each Contracting Party, as far as possible and as appropriate, shall:*

(a) *Introduce appropriate procedures requiring environmental impact assessment of its proposed projects that are likely to have significant adverse effects on biological diversity with a view to avoiding or minimizing such effects and, where appropriate, allow for public participation in such procedures;*

(b) *Introduce appropriate arrangements to ensure that the environmental consequences of its programmes and policies that are likely to have significant adverse impacts on biological diversity are duly taken into account;*

(c) *Promote, on the basis of reciprocity, notification, exchange of information and consultation on activities under their jurisdiction or control which are likely to significantly affect adversely the biological diversity of other states or areas beyond the limits of national jurisdiction, by encouraging the conclusion of bilateral, regional or multilateral arrangements, as appropriate;*

(d) *In the case of imminent or grave danger or damage, originating under its jurisdiction or control, to biological diversity within the area under jurisdiction of other states or in areas beyond the limits of national jurisdiction, notify immediately the potentially affected states of such danger or damage, as well as initiate action to prevent or minimize such danger or damage; and*

(e) *Promote national arrangements for emergency responses to activities or events, whether caused naturally or otherwise, which present a grave and imminent danger to biological diversity and encourage international cooperation to supplement such national efforts and, where appropriate and agreed by the states or regional economic integration organizations concerned, to establish joint contingency plans.*

2. *The Conference of the parties shall examine, on the basis of studies to be carried out, the issue of liability and redress, including restoration and compensation, for damage to biological diversity, except where such liability is a purely internal matter.*

The formal approach encompassed the, *modus operandi*, as defined by statute with guidance with '*when* and *how*', preparation of ER (Environmental Reports) for specified projects, consideration of EA (Environmental Assessment) for decision making, consultative meeting with the public for arriving at a decision, avenues for appeal against official verdicts and apparent irregularities, concerns of integrity having little chance of political impact and supportive media, accessibility to the courts and legal rights of concerning organization. The skilled and vigilant public can play a most significant role in environmental impact assessment (EIA) and environmental conservation.

The Conference of the Parties in the meetings of stressed the need for the development of guidelines for incorporating biodiversity-related issues into environmental-impact- assessment legislation or processes and strategic impact assessment through collaboration with relevant organizations. Contributions on this subject are aimed at different aspects such as Capacity Building and Impact Assessment (CBIA), electronic discussion forum on relevant indicators, impact of trade on biodiversity (www.UNEP/CBD/COP7/1./18.4.04 & www.UNEP/CBD/SBSTTA/10/3 p 39).

In recent years concern over the long-term negative impacts of developmental activities on sacred sites and lands and waters traditionally occupied or used by indigenous and local communities, and in particular the loss of the knowledge, innovations and practices of indigenous and local conservation embodying traditional lifestyles relevant for the conservation and sustainable use of biodiversity (Article 8 j of CBD) have been expressed (www.UNEP/CBD/COP/7/CRP3 p 7). It is emphasized that the conduct of impact assessment within an integrated process will increase the effectiveness of the involvement of indigenous and local communities. There is emphasis on taking appropriate steps to ensure full transparency of the assessment process proposed developments prior to their implementation and exchange of information among different relevant agencies, developers, and communities. The later has relation to the access and benefit sharing (ABS), an important issue of CBD.

From the points of view of purpose and approach Akwe:Kon Voluntary Guidelines in accordance with Art. 8 j of CBD for the conduct of cultural, environmental and social impact assessments regarding developments proposed to take place on, or which are likely to impact on, sacred sites and on lands and water traditionally occupied or used by indigenous and local communities provide a collaborative framework while considering different agencies, communities, developmental authorities and decision makers. These are as following:

(a) *Support the full and effective participation and involvement of indigenous and local communities in screening, scooping and development planning exercises*;

(b) *Properly take into account the cultural, environmental and social concerns and interests of indigenous and local communities, especially of women who often bear a disproportionately large share of negative development impacts*;

(c) *Take into account the traditional knowledge, innovations and practices of indigenous and local communities as part of environmental, social and cultural impact – assessment processes, with due regard to the ownership of and the need for the protection and safe guarding of traditional knowledge, innovations and practices;*

(d) *Promote the use of appropriate technologies*;

(e) *Identify and implement appropriate measures to prevent or mitigate any negative impacts of proposed developments,*

(f) *Take into consideration the interrelationships among cultural, environmental and social elements.*

Need for the Assessment of Environmental Impact

The concept of EIA is known since the year 1969 when US National Environmental Policy Act (NEPA) incorporated the need for the assessment of environmental impact of 'major federal actions significantly affecting the quality of the human environment'. Following is the text of NEPA (Section 2) concerning EIA duly amended in 1986 through a new legislation:

> "*...to declare a national policy which will encourage production and enjoyable harmony between man and his environment to promote efforts which will prevent or eliminate damage to the environment and biosphere and stimulate health and welfare of man, to enrich the understanding of the ecological systems and natural resource important to the Nation and to establish a Council on Environmental Quality*".

The legislation was enacted by the US Congress into a "National Biodiversity Act" for the maintenance of diversity as a national goal, establish mechanisms for coordinating diversity,and set priorities for diversity conservation, and set priorities for diversity conservation. Subsequently in the year 1988 a bill on "National Biological Diversity Conservation and Environmental Research Act" was introduced. The bill proposed several changes including an amendment to NEPA by including phrase "including biological diversity". The bill proposed to change the term "EIA" to include any impact on biological diversity. Further the term, "diversity" was considered to connote "biological diversity" out of "an environment which support diversity and variety of individual choice".

Campbell (1993) believed that if environmental concerns are to become an integral part of the policy making and development planning process in developing countries, then further development by the concerned parties is called for in the procedures, techniques, mode of application, arrangement with capacity building programmes of EIA. He finds following opportunities for further development of EIA and its application which arise from:

- the use to which EIA is put - as a watchdog for negative impacts, rather than a tool for developing positive impacts;
- the limitations of EIA inherent in its use for adjusting pre-designed projects;
- the increasingly technical, esoteric and "stand alone" character of EIA, which is contributing to a disturbing trend away from integration in development planning;
- the incidental, rather than central, role which socio-ecological aspects play in the process and application of EIA;
- the widening gap, in developing countries, between the environmentalist and the decision-maker, and between EIA theory and practice.

The purpose of EIA is in fact to provide the environment its due place in the decision-making process by clearly evaluating the environmental consequences of a proposed activity before action is taken. The concept of EIA has ramifications in the long run for almost all development activity because sustainable development depends on protecting the natural resources which is the foundation for further development.

Status of EIA Process in India and other countries

EIA process is of recent origin in India with emphasis largely on qualification approaches for measurement of the level of pollution of air and water. It deserves a mention that until the government clears the projects for the development of hydro power stations, ports and harbors, thermal power stations and mining, the activities can not begin. In certain development projects adverse impacts on the biodiversity emerge later and indirectly, and therefore, it cannot be presumed or adjudged at early stage of assessment. For example, the oil refineries smoking gas fire burn the insect pollinators, *en masse* resulting into the loss of biological diversity through poor pollination and production.

Voluntary guidelines of Akwe: Kon (UNEP/CBI)/COP/7) for the conduct of cultural environmental and social impact assessment regarding a development proposed to take place on, or which is likely to impact on sacred sites and on lands and waters traditionally occupied or used by indigenous and local communities the guidelines for incorporating biodiversity – related issues into EIA legislation and/or process and in strategic EA, should be taken into account. It is mentioned in the guidelines that EIA can contribute to the protection of the rights of indigenous and local communities by recognizing the distinct activities, customs and beliefs of the affected indigenous and local communities. There is urgent need to assess the direct impacts on local biodiversity at the ecosystem, species and genetic levels, and particularly in terms of those components of biological diversity that the affected indigenous or local community is dependent upon for livelihood. Introduction of invasive species should also be assessed.

There is need to undertake baseline studies on the components of biodiversity of particular significance to the affected indigenous or local community. The studies should include knowledge and valuation of biological resources. Following information should be gathered through studies:

(a) Species inventories

(b) Identification of threatened species with reference to CITES, IUCN, Red List and national inventories.

(c) Identification of specifically significant habitats (relict vegetation, corridors, buffer zones, breeding grounds etc.) and crucial breeding seasons for endangered and critical species.

(d) Identification of areas of particular economic significance (fishing grounds, grazing lands, hunting areas, harvesting areas etc.).

(e) Identification of specifically significant physical features and other natural factors of biodiversity and ecosystems relevance (springs, water courses, lakes, wetlands etc.).

(f) Identification of sites of sacred, religious significance (sacred groves, totemic sites/heritage sites).

Countries with EIA Legislation

Following countries rich in biodiversity and genetic resources have [legislation/no]tification with respect to the protection of their environment.

Table showing countries and the year of EIA notification.

Country	Year
Australia	1974
Brazil	1986
China	1979
Colombia	1974
India	1986
(Environment Protection Act No.29. & EIA Notification, 1994)	1986
Indonesia	1982
Japan	1972
Korea (South)	1977,1979,1981, 1982,1983 & 1986
Malaysia	1985
Mexico	1980, 1982
New Zealand	1991
Pakistan	1983
Papua New Guinea	1973
Philippines	1977
Singapore	1972
Sri Lanka	1984, 1986
Taiwan	1979
Thailand	1975,1978
Venezuela	1976
USA	1969

Eco-sensitive/Fragile Areas

The Indian Environmental (Protection) Act, 1986 notifies following ecologically sensitive/fragile areas in the country:

* Forest

- Key areas of conservation (Wildlife Sanctuaries, National Park

- Biosphere Reserves
- Wetlands
- Coastal Areas
- Aravalis
- Siwaliks
- Western Ghats
- World Heritage sites
- Mangroves
- Hilly areas
- Cold Deserts
- Reserved and Protected Forests
- Closed Areas
- Forest Corridors offering connectivity to critical wildlife habitats
- Islands of critical habitats
- *Ex situ* conservation areas

** Doon Valley, Murud-Janjira, Dahura Talika, Aravali ranges of Gurgaon (Haryana), Alwar District (Rajasthan) (Soni, 2002).

*** Coastal Regulation Zone Notification identifies coastal areas of India under environmental subject.

Environmental impacts *vis-à-vis* monitoring, evaluation and conservation of biological diversity are in the form of following.

1. Ecological and pollution effects on

- Air, water, noise, light and sound.
- Biological diversity – flora, fauna, ecosystem, food web, domestic animal diversity ecosystem.
- Soil erosion, drainage, land degradation.
- Landscapes (artificial and natural)
- Heritage sites
- Drainage and sewerage, waste generation.
- Climate

2. **Natural resource effects on**

- Agric biodiversity
- Forest resource and biodiversity
- Marine and coastal/mangrove resource
- Energy resource (solar, wind, water etc.)
- Construction materials.

3. **Social effects on**

- Community facilities.
- Indigenous communities
- Socio-economic profile of the affected community.
- Recreational facilities
- Settlement aspects (land use, housing, welfare etc.)
- Employment

4. **Economic aspects effect on**

- Rural, urban and ethnic infrastructure development
- Migration of rural community to urbanized world for job or settlement.
- Discrimination on equitable/benefit sharing IPR related issues on the products of biological and environmental origin.

5. **Biosafety – An EIA tool**

The term is defined as "*the policies and procedures adopted to ensure the environmentally safe application of modern biotechnology*" (Persley *et al.* 1992). Biosafety concerns with the release of organism for its nature, locality of release and possibility of subsequent interaction with other elements. This ensures the environmentally safe application of modern biotechnology evolving GMOs (genetically modified organisms). GMOs have following environmental impacts:

- Direct (biological, ecological and evolutionary impacts, or 'biosafety' *per se*).
- Indirect (consequences of adopting a new technology).

- Secondary (sociological and socio-economic)

On one end GMO plays a role of support to conservation and sustainable use of genetic resource while on other hand it posses potential environmental damage through its dominance by fecundity, hybridization with indigenous species and competition with latter. It is further asserted here that safety assessment should be based on the nature of the organism and the environment into which it would be introduced and not on the method by which it was modified (NAS, 1987). Most of the developing countries except India, Mexico and Philippines due to weak enforcement system, lack of financial support and inadequate institutional capacity do not have monitoring and regulatory measures (Krattiger and Lesser, 1995). Most of industrialized countries of OECD (Organization for Economic Co-operation and Development) have biosafety regulations.

The reference to this aspect (Biosafety) may be made to the Convention on Biological Diversity (CBD) under the following Articles.

Article 8 (g) – Parties to the convention should establish biosafety regulatory procedures.

"Establish or maintain means to regulate, manage or control the risks associated with the use and release of living modified organisms resulting from biotechnolgy which are likely to have adverse environmental impacts that could affect the conservation and sustainable use of biological diversity, taking also into account the risks to human health."

Art. 19.3 – Parties to consider the need for, and modalities of, a protocol in the field of the safe transfer, handling and use of GMOs.

"The parties shall consider the need for and modalities of protocol setting out appropriate procedures, including, in particular, advance informed agreement, in the field of safe transfer, handling and use of any living modified organism resulting from biotechnology that may have adverse effect on the conservation and sustainable use of biological diversity."

IUCN (1993), UNEP (1993) interpreted the Articles 6(b), 7(c), 7(h), 14.1 (a), 14.1 (c), 14.1 (d), 17.1 and 18.3 with respect to biosafety to the potential extent.

Following organizations provide assistance in developing appropriate regulations and technical expertise for implementing biosafety rules.

IUCNDO/ICGEB (United Nations Industrial Development Organization/International Centre for Genetic Engineering and Biotechnology).

WHO (World Health Organization).

FAO (Food and Agriculture Organization)

APHIS (Animal and Health Inspection Service) of USA's Department of Agriculture.

ISAAA (International Service for the Acquisition of Biotech Applications).

ISNAR and IBS (International Service for National Agriculture Research) and the Intermediate Biotechnology Service.

USAID (United States Agency for International Development). Others

6. Climate change – a cause of environmental impact on the biodiversity conservation

By the end of this century, global mean surface temperature is expected to increase between 1.5 °C and 6 °C. Sea levels are projected to rise by 15 to 95 cm. The expected warming is greatest over the interior of semi-arid margins of the Sahara and central southern Africa. It is indicated that temperature rise in Africa corresponds to global temperature rise, and that adverse impacts including extremes, are spread across the diverse environments of Africa, putting a huge proportion of African continent at great risk.

Devoy (1987) believes that increase in temperature between 1-5 degree celsius due to greenhouse effect would result into sea level rise through the global warming. UNEP estimated an increase of temperature by 1.5 degree celsius and a sea level rise of 20 cm by 2025. This would lead to following:

- Submergence of forest biodiversity.
    * Coastal areas
    * Low lying areas
    - Impact on agric-biodiversity.

Alter differentially the regeneration through change in microclimate per microhabitat.

Break in paradigm of species composition at all levels.

Impact on the growth and mortality rates of tree species.

Probabilities and rates of wildfire spread due to change in moisture regime.

Fecundity or low reproduction of insect fauna (for examples, *Gigasipham macrosiphon* (Harms) Brenan known from 4 localities in Tanzania and Kenya; population of teak defoliator (*Hyblea puera*) increasing during the hottest period of the year gradually declining with the onset of monsoons. Simultaneously fecundity of teak skeletonizer (*Eutectona machacralis*) results into heavy defoliation. During 1983-84 Lolab Valley in Jammu and Kashmir witnessed the epidemic due to an unusual drought in the region during previous years (1980-83). This seems to be due to the fact that the natural regulating mechanism of the defoliator was completely wiped off (*see* Tewari 1994). Beeson (1941) mentions that the population of *Shorea robusta* (Sal) heartwood borer (*Hoplocerambyx spinicornis*) reaches to the peak of graph during the rainy season with more rainy (monsoon) days but falls with the late rains or no raining days. Yet another example is that of poplar defoliators (*Pygaera cupreata* and *P. fulgurita*) whose population touches height in September with favourable humidity and temperature. Open aspects coupled with climatic factors of different species compared to the shadier one invites diseases and pest, more particularly among the nurseries of plants typical of all together different locality factors of the natural area of occurrence, invasion of weeds and alien species may also be the result of change in micro- climate. Conservation and management of mangrove biodiversity can be considered to certain extent as the components which minimized the impacts of Tsunami waves that occurred along the coasts of Andaman and Nicobar Islands and southern coasts of India on 26th December 2004.

Mangrove ecosystems combat not only the environmental hazards as they contain a rich genetic diversity including tolerance to sea water, sink to the surface and air particulate but also in breaking the cyclone impacts of heavy winds and tides.

From local to international levels (global), micro to megaclimatic and biodiversity levels, and impacting pressure on the lives of citizens environmental impacts can further be categorized into mega environment and micro environmental problems affecting directly and indirectly the biological diversity and subsequently in developing the management strategies for their conservation.

Need and Priorities for the EIA Integrated Biodiversity Conservation:

Following are the main aspects for the development of need and priorities through research, education and extension.

- Characterization and prioritization of environmentally sensitive and threatened species and sites of biodiversity significance applying integrated recent techniques and using RS, GPS and GIS respectively.
- Development of environment and species and GMOs specific biosafety.
- Institutional capacity building in developing countries biosafety, socio-economics, community based social efforts.
- Development of biodiversity conservation – oriented regulations legal issues and policy with respect to the environmental impacts of the components such as GMOs, introduction of biological invasive consciously/subconsciously.
- Involvement of indigenous communities affected or to be affected for the ecological/economic/environmental-cum-urban forestry program impact assessment where their subjects are concerned most.
- National and State level Biodiversity Authority and Biodiversity Management Committees are needed to include all aspects of environment impact assessment of biodiversity conservation.

- Need for the improvement of 'procedures, techniques, mode of application, institutional cum-interdisciplinary research support and networking.
- Setting the aspects of studies on reassessments of the sites/ components studied positive or negative.
- Development of database integrating the significant components of EIA-cum-biodiversity conservation.
- Developing artificial intelligence for environmental-cum-biodiversity modeling. This should include the aspects of early warning system to manage the disaster and crises in foreseeable future.

Checklist on Biodiversity Impact Assessment Interface with Environmental, Ecological, Developmental and Economic Aspects

1. *Proposed site under potential impact*
2. *Components of site*
    - Physical features
    - Geological features
    - Flora
    - Fauna
    - Species inventories
    - Landscapes
    - Phytogeographical
    - Zoogeographical
    - Ecosystem specific
    - Mountain
    - Desert
    - Forest
    - Coastal
    - Wetland
    - Marine

- Mangrove
- Modified landscape
- Ecotone
- Soil (under ground micro flora & fauna)
- Environmentally stressed sites (under the severe impacts of air, noise, light, nuclear, burning of refinery gases).
- Urban (Forestry).
- Rural (Forestry)
- Ethnic (Community interdependent).

3. *Significant features of the area*
   - Land races
   - Threatened habitats of biological importance (Microhabitat)
   - Noble trees (largest/keystone/flagship/umbrella)
   - Rare and Threatened species.

4. *History of the site* (Significance with time & space)
   - Biological
   - Pedological
   - Economic
   - Environmental (climatic)
   - Demographic
   - Political
   - Policies
   - Ethnic (cultural, heritage items etc.)
   - Pasture land history *vis-à-vis* migratory routes.

5. *Identification of rare and threatened species*
   - CITES

PLATE V

Some Threatened Wild Animals of India

1. Swamp Deer
2. One Horned Rhino
3. Golden Langur
4. Flying Squirrel

Photo: By Debojyoti

- IUCN Red Data Book
- National inventories/Reports

6. *Significant habitat*
   - Sacred groves
   - Relict vegetation/site
   - Corridors
   - Refugia
   - Buffer zones
   - Breeding grounds/pool
   - Migratory routes

7. *Physical features and other natural factors of biodiversity significance*
   - Springs
   - Water courses
   - Lakes
   - Wetlands
   - Quarries/mines

8. *Species of economic/cultural importance* (to the communities)
   - Food
   - Fodder
   - Medicine
   - Fuel
   - Construction
   - Fibre
   - Veneration
   - Venomous species
   - Pest or nuisance population (Gilpin 1995)

- Species sensitive to the local environment for proliferation/fecundity.

9. *Areas of economic significance*
   - Fishing grounds
   - Grazing lands
   - Hunting area
   - Harvesting areas

10. *Biological elements of environmental hazards and safety points of view*
    - Invasive alien species and weeds
    - Allergenic plants
    - Poisonous species

11. *Possible effects* (impacts) *of the developmental activities* (Short or Long term)
    - Terrestrial flora & fauna
    - Aquatic flora and fauna
    - Marine biological diversity
    - Coastal flora and fauna
    - Desert flora and fauna
    - Soil micro flora and fauna
    - Avifauna
    - Subterranean/Arboreal/Epiphytic flora and fauna
    - Geological formation sensitive to seismic activity
    - Ecotone
    - Ground Water Table
    - Atmospheric impacts (air/noise/light)
    - Agric ecosystem
    - Village ecosystem with nondescript livestock

- Synthetic fertilizers
- Pesticides (cyclic to nature)
- HYV replacing local varieties
- Erosion of local knowledge and interdependence risks
- Synergism
- Trans boundary effects including corridors and disjuncts
- Man and Animal Conflict

2. *Mitigation of effects*
   - Ex situ conservation
   - Selection of alternate site
   - Relocation
   - Reintroduction
   - Technical measure
   - Financial measure.

3. *Artificial developments of the site*
   - Mega Highways
   - Roads/Streets
   - Railways
   - Air ways (network)
   - Agricultural activities
   - Sub-urban/rural transformation

4. *Viable alternative for the development*
   - Alternative designs or techniques
   - Less ecological risks.
   - Degree of adjustability
   - Compensation to the holders.
   - Preservation of ecotone or site

- Benefits
- Relocation-prospects and constraints.

15. *Implications*
    - Effects of road building
    - Drainage of wet areas
    - Dumps of geological features
    - Submergence of site for dams
    - Clear felling
    - Selected systems or operation of timber extraction
    - Local varieties replacement
    - Forest fragmentation causing genetic isolation of flora/fauna
    - Over exploitation/harvesting
    - Mining
    - Illegal trade
    - Surface mines
    - Coral for cement lime etc.
    - Pollutants (different ecosystems)
    - Air (industrial, burgery)
    - Water (sewage/waste)
    - Encroachment/displacement
    - Migration of population to newer and congenial environment

International and National EIA Agencies

Some of important agencies which can contribute towards th development of biological diversity with respect to environmental, ecologica and economic aspects are mentioned as following:

- Environment and Economics Unit (UNEP), Nairobi, Kenya
- Infoterra (Global Environmental Information Exchange Networ of UNEP World Bank, Washington DC. 20433, USA.

- EIA Centre, Manchester, UK.
- Institute for European Environmental Policy (IEEP), London, UK.
- International Association for Impact Assessment (IAIA), Fargo, USA
- International Institute for Environment and Development (IIED), London, United Kingdom.
- World Conservation Union (IUCN), Geneva, Switzerland.
- Impact Assessment Centre, Ontario, Canada
- UN Economic and Social Council for Asia and the Pacific (ESCAP), Bangkok Thailand.
- International Environmental Technology Centre (UNEP), Osaka, Japan.
- Bureau for Coordination of Environmental Program (UNESCO), Paris, France.
- Food and Agriculture Organization (FAO), Rome, Italy.
- Centre for Environmental Science & Engineering (CESE), IIT, Mumbai, India.
- Ghose Bose and Associates Pvt. Ltd. Kolkata, West Bengal, India.
- Indian Institute of Ecology and Environment, New Delhi, India.
- National Environment Engineering Research Institute (NEERI), Nagpur, India
- Ministry of Environment & Forests, Lodi Road, New Delhi, India
- Central Pollution Control Board, East Arjun Nagar, New Delhi, India.
- Centre for Science & Environment, New Delhi, India.
- Kalpavriksh – Environmental Action Group, New Delhi, India.
- Tata Energy Research Centre (TERI), New Delhi, India.
- Society for Technology and Action for Rural Advancement (TARA), New Delhi, India.
- Environmental Protection Research Foundation (EPRF), Sangli, India.

- World Wide Fund for Native India, New Delhi and Mumbai, India.
- Zoo Outreach Organization (ZOO) Mysore, Karnataka, India.
- Centre for Ecological Science, Banglore, India.
- Bombay Natural History Society (BNHS), Mumbai, India.
- Peoples Institute for Participatory Action Research (PIPAR), Orissa, India.
- Madras Crocodile Bank Trust (MCBT), Chennai, Tamil Nadu, India.
- Peoples Science Institute (PSI) Dehra Dun, Uttaranchal, India
- Botanical Survey of India, Kolkata, West Bengal, India.
- Zoological Survey of India, Kolkata, West Bengal, India.
- Salim Ali Centre for Ornithology and National History, Coimbatore, India Institute for Forest Management (IIFM), Bhopal, M.P., India.
- Gujarat Institute of Desert Ecology (GUIDE), Bhuj, Gujarat, India.
- Indian Institute of Remote Sensing (IIRS), Dehra Dun, India.
- National Remote Sensing Agency, Hyderabad, India.
- Wildlife Institute of India, Dehra Dun, Uttaranchal, India.
- Central Water Commission, New Delhi, India.
- Indian Council of Forestry Research & Education (ICFRE) and Institutes, Dehra Dun, Uttaranchal, India.

Literature Cited

Beeson, CFC (1941): The ecology and control of forest insects of India and neighboring countries, Vasant Press, Dehra Dun.

Campbell, Ian (1993):: Environmental Impact. Where to from here? For discussion at the UNEP Consultation Meeting "The Future of EIA in developing Countries". Paris, 27-28th October 1993. Environmental Economic Series Paper No. 6. :1-2.

Devoy, RJN (1987): Sea surface studies, a global view. Croomhelm. London.

Gilpin, A. (1995) Environmental impact assessment – cutting edge for the twenty first century. Cambridge University Press, UK.

IUCN (1993): The Convention on Biological Diversity – an explanatory guide – Draft Text. IUCN – The World Conservation Union, Gland.

Krattiger, AF and Lesser, WH (1995): Biosafety – an environmental impact assessment tool – and the role of the Convention on Biological Diversity. Widening Perspectum in Biodiversity. *In*: Course Director, Biodiversity Conservation Course for SFS Officers 15-26th November 1999: 353-366 State Forest Service College, Dehra Dun.

NAS (National Academy of Sciences) (1987): Introduction of recombinant DNA – engineered organisms into the environment: Key issues, NAS, National Academic Press, Washington DC.

Persley, GJ, Giddings, LU and Juma, C. (1992): Biosafety – the safe application of biotechnology in agriculture and the environment. International Service for National Agricultural Research. The Hague., Netherlands.

Soni, P.L. (2002): Ecological impact assessment and restoration of biodiversity in mine areas. *In*: Lecture notes on Short Term Training Course on Biodiversity Conservation organized by FRI. Dehra Dun :142.

Tewari, DN (1994): Forests and environment: 293-296. International Book Distributors, Dehra Dun.

UNEP (1993): Expert Panel Meeting on Biological Diversity. Final Report of Expert Panel 4. United Nations Environment Program (UNEP), Nairobi.

CONVENTION ON BIOLOGICAL DIVERSITY - TEXT AND ANNEXES

Introduction

The earth's biological resources are vital to humanity's economic and social development. As a result, there is a growing recognition that biological diversity is a global asset of tremendous value to present and future generations. At the same time, the threat to species and ecosystems has never been so great as it is today. Species extinction caused by human activities continues at an alarming rate.

In response, the United Nations Environment Programme (UNEP) convened the Ad Hoc Working Group of Experts on Biological Diversity in November 1988 to explore the need for an international convention on biological diversity. Soon after, in May 1989, it established the Ad Hoc Working Group of Technical and Legal Experts to prepare an international legal instrument for the conservation and sustainable use of biological diversity. The experts were to take into account "the need to share costs and benefits between developed and developing countries" as well as "ways and means to support innovation by local people"

By February 1991, the Ad Hoc Working Group had become known as the Intergovernmental Negotiating Committee. Its work culminated on 22 May 1992 with the Nairobi Conference for the Adoption of the Agreed Text of the Convention on Biological Diversity.

The Convention was opened for signature on 5 June 1992 at the United Nations Conference on Environment and Development (the Rio "Earth Summit"). It remained open for signature until 4 June 1993, by which time it had received 168 signatures. The Convention entered into force on 29 December 1993, which was 90 days after the 30th ratification. The first session of the Conference of the Parties was scheduled for 28 November 9 December 1994 in the Bahamas.

"The Convention on Biological Diversity was inspired by the world community's growing commitment to sustainable development. It represents a dramatic step forward in the conservation of biological diversity, the sustainable use of its components, and the fair and equitable sharing of benefits arising from the use of genetic resources.

Convention on Biological Diversity - Preamble

The Contracting Parties,

Conscious of the intrinsic value of biological diversity and of the ecological, genetic, social, economic, scientific, educational, cultural, recreational and aesthetic values of biological diversity and its components.

Conscious also of the importance of biological diversity for evolution and for maintaining life sustaining systems of the biosphere.

Affirming that the conservation of biological diversity is a common concern of humankind.

Reaffirming also that States are responsible for conservation their biological diversity and for using their biological resources in a sustainable manner.

Concerned that biological diversity is being significantly reduced by certain human activities.

Aware of the general lack of information and knowledge regarding biological diversity and of the urgent need to develop scientific, technical and institutional capacities to provide the basic understanding upon which to plan and implement appropriate measures.

Noting also that where there is a threat of significant reduction or loss of biological diversity, lack of full scientific certainty should not be used as a reason for postponing measures to avoid or minimize such a threat.

Noting further that the fundamental requirement for the conservation of biological diversity is the *in-situ* conservation of ecosystems and natural habitats and the maintenance and recovery of viable populations of species in their natural surroundings.

Noting further that *ex-situ* measures, preferably in the country of origin, also have an important role to play.

Recognizing the close and traditional dependence of many indigenous and local communities embodying traditional lifestyles on biological resources, of the desirability of sharing equitably benefits arising from the use of traditional knowledge, innovations and practices relevant to the observation of biological diversity and the sustainable use of its components.

Recognizing also the vital role that women play in the conservation and sustainable use of biological diversity and affirming the need for the full participation of women at all levels of policy-making and implementation for biological diversity conservation.

Stressing the importance of, and the need to promote, international, regional and global cooperation among States and intergovernmental organizations and the non-governmental sector for the conservation of biological diversity and the sustainable use of its components.

Acknowledging that the provision of new and additional financial resources and appropriate access to relevant technologies can be expected to make a substantial difference in the world's ability to address the loss of biological diversity.

Acknowledging further that special provision is required to meet the needs of developing countries, including the provision of new and additional financial resources and appropriate access to relevant technologies.

Noting in this regard the special conditions of the least developed countries and small island States.

Acknowledging that substantial investments are required to conserve biological diversity and that there is the expectation of a broad range of environmental, economic and social benefits from those investments.

Recognizing that economic and social development and poverty eradication are the first and overriding priorities of developing countries.

Aware that conservation and sustainable use of biological diversity is of critical importance for meeting the food, health and other needs of the growing world population, for which purpose access to and sharing of both genetic resources and technologies are essential.

Noting that, ultimately, the conservation and sustainable use of biological diversity will strengthen friendly relations among States and contribute to peace for humankind.

Desiring to enhance and complement existing international arrangements for the conservation of biological diversity and sustainable use of its components, and

Determined to conserve and sustainably use biological diversity for the benefit of present and future generations.

We agreed as follows:

Article 1

Objectives

The objectives of this Convention, to be pursued in accordance with its relevant provisions, are the conservation of biological diversity, the sustainable use of its components and the fair and equitable sharing of the benefits arising out of the utilization of genetic resources, including by appropriate access to genetic resources and by appropriate transfer of relevant technologies, taking into account all rights over those resources and to technologies, and by appropriate funding.

Article 2

Use of Terms

For the purposes of this Convention:

"*Biological diversity*" means the variability among living organisms from sources including, *inter alia*, terrestrial, marine and other aquatic ecosystems and the ecological complexes of which they are part; this includes diversity within species, between species and of ecosystems.

"Biological resources" includes genetic resources, organisms or parts thereof, populations, or any other biotic component of ecosystems with actual or potential use or value for humanity.

"Biotechnology" means any technological application that uses biological systems, living organisms, or derivatives thereof, to make or modify products or processes for specific use.

"Country of origin of genetic resources" means the country which possesses those genetic resources in *in-situ* conditions.

"*Country providing genetic resources"* means the country supplying genetic resources collected from *in-situ* sources, including populations of both wild and domesticated species, or taken from *ex-situ* sources, which may or may not have originated in that country.

"*Domesticated or cultivated species"* means species in which the evolutionary process has been influenced by humans to meet their needs.

"*Ecosystem"* means a dynamic complex of plant, animal and micro-organism communities and their non-living environment interacting as a functional unit.

"*Ex-situ conservation"* means the conservation of components o biological diversity outside their natural habitats.

"*Genetic material"* means any material of plant, animal, microbial or other origin containing functional units of heredity.

"*Genetic resources"* means genetic material of actual or potential value.

"*Habitat"* means the place or type of site where an organism or population naturally occurs.

"*In-situ conditions*" means conditions where genetic resources exist within ecosystems and natural habitats, and, in the case of domesticated or cultivated species, in the surroundings where they have developed their distinctive properties.

"*In-situ conservation*" means the conservation of ecosystems and natural habitats and the maintenance and recovery of viable populations of species in their natural surroundings and, in the case of domesticated or cultivated species, in the surroundings where they have developed their distinctive properties.

"*Protected area*" means a geographically defined area which is designated or regulated and managed to achieve specific conservation objectives.

"*Regional economic integration organization*" means an organization constituted by sovereign States of a given region, to which its member States have transferred competence in respect of matters governed by this Convention and which has been duly authorized, in accordance with its internal procedures, to sign, ratify, accept, approve or accede to it.

'*Sustainable use*" means the use of components of biological diversity in a way and at a rate that does not lead to the long-term decline of biological diversity, thereby maintaining its potential to meet the needs and aspirations of present and future generations.

"*Technology*" includes biotechnology.

Article 3

Principle

States have, in accordance with the Charter of the United Nations and the principles of international law, the sovereign right to exploit their own resources pursuant to their own environmental policies, and the responsibility to ensure that activities within their jurisdiction or control do not cause damage to the environment of other States or of areas beyond the limits of national jurisdiction.

Article 4

Jurisdictional Scope

Subject to the rights of other States, and except as otherwise expressly provided in this Convention, the provisions of this Convention apply, in relation to each Contracting Party:

(a) In the case of components of biological diversity, in areas within the limits of its national jurisdiction; and

(b) In the case of processes and activities, regardless of where their effects occur, carried out under its jurisdiction or control, within the area of its national jurisdiction or beyond the limits of national jurisdiction.

Article 5

Cooperation

Each Contracting Party shall, as far as possible and as appropriate, operate with other Contracting Parties, directly or, where appropriate, rough competent international organizations, in respect of areas beyond national jurisdiction and on their matters of mutual interest for the conservation and sustainable use of biological diversity.

Article 6

General Measures for Conservation and Sustainable Use

Each Contracting Party shall, in accordance with its particular conditions and capabilities:

(a) Develop national strategies, plans or programmes for the conservation and sustainable use of biological diversity or adapt for this purpose existing strategies, plans or programmes which shall reflect, *inter alia*, the measures set out in this Convention relevant to the Contracting Party concerned; and

(b) Integrate, as far as possible and as appropriate, the conservation and sustainable use of biological diversity into relevant sectoral or cross-sectoral plans, programmes and policies.

Article 7

Identification and Monitoring

Each Contracting Party shall, as far as possible and as appropriate, in particular for the purpose of Articles 8 to 10:

(a) Identify components of biological diversity important for its conservation and sustainable use having regard to the indicative list of categories set down in Annex I;

(b) Monitor, through sampling and other techniques, the components of biological diversity identified pursuant to subparagraph (a) above, paying particular attention to those requiring urgent conservation measures and those which offer the greatest potential for sustainable use;

(c) Identify processes and categories of activities which have or are likely to have significant adverse impacts on the conservation and sustainable use of biological diversity, and monitor their effects through sampling and other techniques; and

(d) Maintain and organize, by any mechanism data, derived from identification and monitoring activities pursuant to subparagraphs (a), (b) and (c) above.

Article 8

In-situ Conservation

Each Contracting Party shall, as far as possible and as appropriate:

(a) Establish a system of protected areas or areas where special measures need to be taken to conserve biological diversity;

(b) Develop, where necessary, guidelines for the selection, establishment and management of protected areas of areas where special measures need to be taken to conserve biological diversity;

(c) Regulate or manage biological resources important for the conservation of biological diversity whether within or outside protected areas, with a view to ensuring their conservation and sustainable use;

(d) Promote the protection of ecosystems, natural habitats and the maintenance of viable populations of species in natural surroundings.

(e) Promote environmentally sound and sustainable development in areas adjacent to protected areas with a view to furthering protection of these areas;

(f) Rehabilitate and restore degraded ecosystems and promote the recovery of threatened species *inter alia*, through the development and implementation of plans or other management strategies;

(g) Establish or maintain means to regulate, manage or control the risks associated with the use and release of living modified organisms resulting from biotechnology which are likely to have adverse environmental impacts that could affect the conservation and sustainable use of biological diversity, taking also into account the risks to human health.

(h) Prevent the introduction of, control or eradicate those alien species which threaten ecosystems, habitats or species;

(i) Endeavour to provide the conditions needed for compatibility between present uses and the conservation of biological diversity and the sustainable use of its components;

(j) Subject to its national legislation, respect, preserve and maintain knowledge, innovations and practices of indigenous and local communities embodying traditional lifestyles relevant for the conservation and sustainable use of biological diversity and promote their wider application with the approval and involvement of the holders of such knowledge, innovations and practices and encourage the equitable sharing of the benefits arising from the utilization of such knowledge, innovations and practices;

(k) Develop or maintain necessary legislation and/or other regulatory provisions for the protection of threatened species and populations;

(l) Where a significant adverse effect on biological diversity has been determined pursuant to Article 7, regulate or manage the relevant processes and categories of activities; and

(m) Cooperate in providing financial and other support for in-situ conservation outlined in subparagraphs (a) to (1) above, particularly to developing countries.

Article 9

Ex-situ Conservation

Each Contracting party shall, as far as possible and as appropriate, and predominantly for the purpose of complementing *in-situ* measures:

(a) Adopt measures for the *ex-situ* conservation of components of biological diversity, preferably in the country of origin of such components;

(b) Establish and maintain facilities for *ex-situ* conservation of and research on plants, animals and micro-organisms, preferably in the country of origin of genetic resources;

(c) Adopt measures for the recovery and rehabilitation of threatened species and for their reintroduction into their natural habitats under appropriate conditions;

(d) Regulate and manage collection of biological resources from natural habitats for *ex-situ* conservation purposes so as not to threaten ecosystems and *in-situ* populations of species, except where special temporary *ex-situ* measures are required under subparagraph (c) above; and

(e) Cooperate in providing financial and other support for *ex-situ* conservation outlined in subparagraphs (a) to (d) above and in the establishment and maintenance of *ex-situ* conservation facilities in developing countries.

Article 10

Sustainable use of Components of Biological Diversity

Each Contracting Party shall, as far as possible and as appropriate:

(a) Integrate consideration of the conservation and sustainable use of biological resources into national decision-making;

(b) Adopt measures relating to the use of biological resources to avoid or minimize adverse impacts on biological diversity;

(c) Protect and encourage customary use of biological resources in accordance with traditional cultural practices that are compatible with conservation or sustainable use requirements;

(d) Support local populations to develop and implement remedial action in degraded areas where biological diversity has been reduced; and

(e) Encourage cooperation between its governmental authorities and its private sector in developing methods for sustainable use of biological resources.

Article 11

Incentive Measures

Each Contracting Party shall, as far as possible and a appropriate, adopt economically and socially sound measures that act as incentives for the conservation and sustainable use of components of biological diversity.

Article 12

Research and Training

The Contracting Parties, taking into account the special needs of developing countries, shall:

(a) Establish and maintain programmes for scientific and technical education and training in measures for the identification, conservation and sustainable use of biological diversity and its components and provide support for such education and training for the specific needs of developing countries;

(b) Promote and encourage research which contributes to the conservation and sustainable use of biological diversity, particularly in developing countries, *inter alia*, in accordance with decisions of the Conference of the Parties taken in consequence of recommendations of the Subsidiary Body on Scientific, Technical and Technological Advice; and

(c) In Keeping with the provisions of Articles 16, 18 and 20, promote and cooperate in the use of scientific advances in biological diversity research in developing methods for conservation and sustainable use of biological resources.

Article 13

Public Education and Awareness

The Contracting Parties shall:

(a) Promote and encourage understanding of the importance of, and the measures required for, the conservation of biological diversity, as well as its propagation through media, and the inclusion of these topics in educational programmes; and

(b) Cooperate, as appropriate, with other States and international organizations in developing educational and public awareness programmes, with respect to conservation and sustainable use of biological diversity.

Article 14

Impact Assessment and Minimising Adverse Impacts

1. Each Contracting Party, as far as possible and as appropriate, shall:

(a) Introduce appropriate procedures requiring environmental impact assessment of its proposed projects that are likely to have significant adverse effects on biological diversity with a view to avoiding or minimizing such effects and, where appropriate, allow for public participation in such procedures;

(b) Introduce appropriate arrangements to ensure that the environmental consequences of its programmes and policies that are likely to have significant adverse impacts on biological diversity are duly taken into account;

(c) Promote, on the basis of reciprocity, notification, exchange of information and consultation on activities under their jurisdiction or control which are likely to significantly affect adversely the biological diversity of other States or areas beyond the limits of national jurisdiction, by encouraging the conclusion of bilateral, regional or multilateral arrangements, as appropriate;

(d) In the case of imminent or grave danger or damage, originating under its jurisdiction or control, to biological diversity within the area under jurisdiction of other States or in areas beyond the limits of national jurisdiction, notify immediately the potentially affected States of such danger or damage, as well as initiate action to prevent or minimize such danger or damge; and

(e) Promote national arrangements for emergency responses to activities or events, whether caused naturally or otherwise, which present a grave and imminent danger to biological diversity and encourage international cooperation to supplement such national efforts and, where appropriate and agreed by the States or regional economic integration organizations concerned, to establish joint contingency plans.

2. The Conference of the Parties shall examine, on the basis of studies on be carried out, the issue of liability and redress, including restoration and compensation, for damage to biological diversity, except where such liability is a purely internal matter.

Article 15

Access to Genetic Resources

1. Recognizing the sovereign rights of States over their natural resources, the authority to determine access to genetic resources rests with the national governments and is subject to national legislation.

2. Each Contracting Party shall endeavour to create conditions to facilitate access to genetic resources for environmentally sound uses by other Contracting Parties and not to impose restrictions that run counter to the objectives of this Convention.

3. For the purpose of this Convention, the genetic resources being provided by a Contracitng Party, as referred to in this Article and Articles 16 and 19, are only those that are provided by Contracting Parties that are countries of origin of such resources or by the Parties that have acquired the genetic resources in accordance with this Convention.

4. Access, where granted, shall be on mutually agreed terms and subject to the provisions of this Article.

5. Access to genetic resources shall be subject to prior informed consent of the Contracting Party providing such resources, unless otherwise determined by that Party.
6. Each Contracting Party shall endeavor to develop and carry out scientific research based on genetic resources provided by other Contracting Parties with the full participation of, and where possible in, such Contracting Parties

7. Each Contracting Party shall take legislative, administrative or policy measures, as appropriate, and in accordance with Articles 16 and 19 and, where necessary, through the financial mechanism established by Articles 20 and 21 with the aim of sharing in a fair and equitable way the results of research and development and the benefits arising from the commercial and other utilization of genetic resources with the Contracting Party providing such resources. Such sharing shall be upon mutually agreed terms.

Article 16

Access to and Transfer of Technology

1. Each Contracting Party, recognizing that technology includes biotechnology, and that both access to and transfer of technology among Contracting Parties are essential elements for the attainment of the objectives of this Convention, undertakes subject to the provisions of this Article to provide and/or facilitate access for and transfer to other Contracting Parties of technologies that are relevant to the conservation and sustainable use of biological diversity or make use of genetic resources and do not cause significant damage to the environment.

2. Access to and transfer of technology referred to in paragraph 1 above to developing countries shall be provided and/or facilitated under fair and most favourable terms, including on concessional and preferential terms where mutually agreed, and, where necessary, in accordance with the financial mechanism established by Articles 20 and 21. In the case of technology subject to patents and other intellectual property rights, such access and transfer shall be provided on terms which recognize and are consistent with the adequate and effective protection of intellectual property rights. The application of this paragraph shall be consistent with paragraphs 3, 4 and 5 below.

3. Each Contracting Party shall take legislative, administrative or policy measures, as appropriate, with the aim that Contracting Parties, in particular those that are developing countries, which provide genetic resources are provided access to and transfer of technology which makes use of those resources, on mutually agreed terms, including technology protected by patents and other intellectual property rights, where necessary, through the provisions of Articles 20 and 21 and in accordance with international law and consistent with paragraphs 4 and 5 below.

4. Each Contracting Party shall take legislative, administrative or policy measures, as appropriate, with the aim that the private sector facilitates access to, joint development and transfer of technology referred to in paragraph 1 above for the benefit of both governmental institutions and

the private sector of developing countries and in this regard shall abide by the obligations included in paragraphs 1, 2 and 3 above.

5. The Contracting Parties, recognizing that patents and other intellectual property rights may have an influence on the implementation of this Convention, shall cooperate in this regard subject to national legislation and international law in order to ensure that such rights are supportive of an do not run counter to its objectives.

Article 17

Exchange of Information

1. The Contracting Parties shall facilitate the exchange of information, from all publicly available sources, relevant to the conservation and sustainable use of biological diversity, taking into account the special needs of developing countries.

2. Such exchange of information shall include exchange of results of technical, scientific and socio-economic research, as well as information on training and surveying programmes, specialized knowledge, indigenous and traditional knowledge as such and in combination with the technologies referred to in Article 16, paragraph 1. It shall also, where feasible, include repatriation of information.

Article 18

Technical and Scientific Cooperation

1. The Contracting Parties shall promote international technical and scientific cooperation in the field of conservation and sustainable use of biological diversity, where necessary, through the appropriate international and national institutions.

2. Each Contracting Party shall promote technical and scientific cooperation with other Contracting Parties, in particular developing countries, in implementing this Convention, *inter alia*, through the development and implementation of national policies. In promoting such

cooperation, special attention should be given to the development and strengthening of national capabilities, by means of human resources development and institution building.

3. The Conference of the Parties, at its first meeting, shall determine how to establish a clearing-house mechanism to promote and facilitate technical and scientific cooperation.

4. The Contracting Parties shall, in accordance with national legislation and policies, encourage and develop methods of cooperation for the development and use of technologies, including indigenous and traditional technologies, in pursuance of the objectives of this Convention. For this purpose, the Contracting Parties shall also promote cooperation in the training of personnel and exchange of experts.

5. The Contracting Parties shall, subject to mutual agreement, promote the establishment of joint research programmes and joint ventures for the development of technologies relevant to the objectives of this Convention.

Article 19

Handling of Biotechnology and Distribution of its Benefits

1. Each Contracting Party shall take legislative, administrative or policy measures, as appropriate, to provide for the effective participation in biotechnological research activities by those Contracting Parties, especially developing countries, which provide the genetic resources for such research, and where feasible in such Contracting Parties.

2. Each Contracting Party shall take all practicable measures to promote and advance priority access on a fair and equitable basis by Contracting Parties, especially developing countries, to the results and benefits arising from biotechnologies based upon genetic resources provided by those Contracting Parties. Such access shall be on mutually agreed terms.

3. The Parties shall consider the need for and modalities of a protocol setting out appropriate procedures, including, in particular, advance informed agreement, in the field of the safe transfer, handling and use

of any living modified organism resulting from biotechnology that may have adverse effect on the conservation and sustainable use of biological diversity.

4. Each Contracting Party shall, directly or by requiring any natural or legal person under its jurisdiction providing the organisms referred to in paragraph 3 above, provide any available information about the use and safety regulations, required by that Contracting Party in handling such organisms, as well as any available information on the potential adverse impact of the specific organisms concerned to the Contracting party into which those organisms are to be introduced.

Article 20

Financial Resources

1. Each Contracting Party undertakes to provide, in accordance with its capabilities, financial support and incentives in respect of those national activities which are intended to achieve the objectives of this Convention, in accordance with its national plans, priorities and programmes.

2. The developed country Parties shall provide new and additional financial resources to enable developing country Parties to meet the agreed full incremental costs to them of implementing measures which fulfill the obligations of this Convention and to benefit from its provisions and which costs are agreed between a developing country Party and the institutional structure referred to in Article 21, in accordance with policy, strategy, programme priorities and eligibility criteria and an indicative list of incremental costs established by the Conference of the Parties. Other parties, including countries undergoing the process of transition to a market economy, may voluntarily assume the obligations of the developed country Parties. For the purpose of this Article, the Conference of the Parties, shall at its first meeting establish a list of developed country Parties and other Parties which voluntarily assume the obligations of the developed country Parties. The Conference of the Parties shall periodically review and it necessary amend the list. Contributions from other countries and sources on a voluntary basis would also be encouraged. The implementation of these commitments

shall take into account the need for adequacy, predictability and timely flow of funds and the importance of burden-sharing among the contributing parties included in the list.

3. The developed country parties may also provide, and developing country Parties avail themselves of, financial resources related to the implementation of this Convention through bilateral, regional and other multilateral channels.

4. The extent to which developing country parties will effectively implement their commitments under this Convention will depend on the effective implementation by developed country Parties of thie commitments under this Convention related to financial resources and transfer of technology and will take fully into account the fact that economic and social development and eradication of poverty are the first and overriding priorities of the developing country Parties.

5. The Parties shall take full account of the specific needs and special situation of least developed countries in their actions with regard to funding and transfer of technology.

6. The Contracting Parties shall also take into consideration the special conditions resulting from the dependence on, distribution and location of, biological diversity within developing country Parties, in particular small island States.

7. Consideration shall also be given to the special situation of developing countries, including those that are most environmentally vulnerable, such as those with arid and semi-arid zones, coastal and mountainous areas.

Article 21

Financial Mechanism

1. There shall be a mechanism for the provision of financial resources to developing country Parties for purposes of this Convention on a grant or concessional basis the essential elements of which are described in this Article. The mechanism shall function under the authority and

guidance of, and be accountable to, the Conference of the Parties for purposes of this Convention. The operations of the mechanism shall be carried out by such institutional structure as may be decided upon by the Conference of the Parties at its first meeting. For purposes of this Convention, the Conference of the Parties shall determine the policy, strategy, programme priorities and eligibility criteria relating to the access to and utilization of such resources. The contributions shall be such as to take into account the need for predictability, adequacy and timely flow of funds referred to in Article 20 in accordance with the amount of resources needed to be decided periodically by the Conference of the Parties and the importance of burden sharing among the contributing Parties included in the list referred to in Article 20, paragraph 2. Voluntary contributions may also be made by the developed country Parties and by other countries and sources. The mechanism shall operate within a democratic and transparent system of governance.

2. Pursuant to the objectives of this Convention, the Conference of the Parties shall at its first meeting determine the policy, strategy and programme priorities, as well as detailed criteria and guidelines for eligibility for access to and utilization of the financial resources including monitoring and evaluation on a regular basis of such utilization. The Conference of the Parties shall decide on the arrangements to give effect to paragraph 1 above after consultation with the institutional structure entrusted with the operation of the financial mechanism.

3. The Conference of the Parties shall review the effectiveness of the mechanism established under this Article, including the criteria and guidelines referred to in paragraph 2 above, not less than two years after the entry into force of this Convention and thereafter on a regular basis. Based on such review, it shall take appropriate action to improve the effectiveness of the mechanism if necessary.

4. The Contracting Parties shall consider strengthening existing financial institutions to provide financial resources for the conservation and sustainable use of biological diversity.

Article 22

Relationship with other International Conventions

1. The provisions of this Convention shall not affect the rights and obligations of any Contracting Party deriving from any existing international agreement, except where the exercise of those rights and obligations would cause a serious damage or threat to biological diversity.

2. Contracting Parties shall implement this Convention with respect to the marine environment consistently with the rights and obligations of State under the law of the sea.

Article 23

Conference of the Parties

1. A Conference of the Parties is hereby established. The first meeting of the Conference of the Parties shall be convened by the Executive Director of the United Nations Environment Programme not later than one year after the entry into force of this Convention. Thereafter, ordinary meetings of the Conference of the Parties shall be held at regular intervals to be determined by the Conference at its first meeting.

2. Extraordinary meetings of the Conference of the Parties shall be held at such other times as may be deemed necessary by the Conference, or at the written request of any Party, provided that, within six months of the request being communicated to them by the Secretariat, it is supported by at least one third of the Parties.

3. The Conference of the Parties shall by consensus agree upon and adopt rules of procedure of itself and for any subsidiary body it may establish, as well as financial rules governing the funding of the Secretariat. At each ordinary meeting, it shall adopt a budget for the financial period until the next ordinary meeting.

4. The Conference of the Parties shall keep under review the implementation of this Convention, and, for this purpose, shall:

(a) Establish the form and the intervals for transmitting the information to be submitted in accordance with Article 26 and consider such information as well as reports submitted by any subsidiary body:

(b) Review scientific, technical and technological advice on biological diversity provided in accordance with Article 25;

(c) Consider and adopt, as required, protocols in accordance with Article 28;

(d) Consider and adopt, as required, in accordance with Articles 29 and 30, amendments to this Convention and its annexes;

(e) Consider amendments to any protocol, as well as to any annexes thereto, and, if so decided, recommend their adoption to the parties to the protocol concerned;

(f) Consider and adopt, as required, in accordance with Article 30, additional annexes to this Convention;

(g) Establish such subsidiary bodies, particularly to provide scientific and technical advice, as are deemed necessary for the implementation of this Convention;

(h) Contact, through the Secretariat, the executive bodies of conventions dealing with matters covered by this Convention with a view to establishing appropriate forms of cooperation with them; and

(i) Consider and undertake any additional action that may be required for the achievement of the purposes of this Convention in the light of experience gained in its operation.

5. The United Nations, its specialized agencies and the International Atomic Energy Agency, as well as any State not Party to this Convention, may be represented as observers at meetings of the Conference of the Parties. Any other body or agency, whether governmental or non-governmental, qualified in fields relating to conservation and sustainable use of biological diversity, which has informed the Secretariat of its wish to be represented as an observer at a meeting of the Conference

of the Parties may be admitted unless at least one third of the Parties present object. The admission and participation of observers shall be subject to the rules of procedure adopted by the Conference of the Parties.

Article 24

Secretariat

1. A secretariat is hereby established. Its functions shall be:

(a) To arrange for and service meetings of the Conference of the Parties provided for in Article 23;

(b) To perform the functions assigned to it by any protocol;

(c) To prepare reports on the execution of its functions under this Convention and present them to the Conference of the parties;

(d) To coordinate with other relevant international bodies and, in particular to enter into such administrative and contractual arrangements as may be required for the effective discharge of its functions; and

(e) To perform such other functions as may be determined by the Conference of the Parties.

2. At its first ordinary meeting, the Conference of the Parties shall designate the secretariat from amongst those existing competent international organizations which have signified their willingness to carry out the secretariat functions under this Convention.

Article 25

Subsidiary body on Scientific, Technical and Technology Advice

1. A subsidiary body for the provision of scientific, technical and technological advice is hereby established to provide the Conference of the Parties and, as appropriate, its other subsidiary bodies with timely advice relating to the

implementation of this Convention. This body shall be open to participation by all Parties and shall be multidisciplinary. It shall comprise government representatives competent in the relevant field of expertise. It shall report regularly to the Conference of the Parties on all aspects of its work.

Article 26

Reports

Each Contracting Party shall, at intervals to be determined by the Conference of the Parties, present to the Conference of the Parties, reports on measures which it has taken for the implementation of the provisions of this Convention and their effectiveness in meeting the objectives of this Convention.

Article 27

Settlement of Disputes

1. In the event of a dispute between Contracting Parties concerning the interpretation or application of this Convention, the parties concerned shall seek solution by negotiation.

2. If the parties concerned cannot reach agreement by negotiation, they may jointly seek the good offices of, or request mediation by, a third party.

3. When ratifying, accepting, approving or acceding to this Convention, or at any time thereafter, a State of regional economic integration organization may declare in writing to the Depositary that for a dispute not resolved in accordance with paragraph 1 or paragraph 2 above, it accepts one or both of the following means of dispute settlement as compulsory.

a) Arbitration in accordance with the procedure laid down in Part I of Annex II;

b) Submission of the dispute to the International Court of Justice.

4. If the parties to the accordance with Part 2 of Annex II unless the parties otherwise agree.

5. The provisions of this Article shall apply with respect to any protocol except as otherwise provided in the protocol concerned.

Article 28

Adoption of Protocols

1. The Contracting Parties shall cooperate in the formulation and adoption of protocols to this Convention.

2. Protocols shall be adopted at a meeting of the Concurrence of the Parties.

3. The text of any proposed protocol shall be communicated to the Contracting Parties, by the Secretariat at least six months before such a meeting.

Article 29

Amendment of the Convention or Protocols

1. Amendments to this Convention may be proposed by any Contracting Party. Amendments to any protocols may be proposed by any Party to that protocol.

2. Amendments to this Convention shall be adopted at a meeting of the Conference of the Parties. Amendments to any protocol shall be adopted at a meeting of the Parties to the Protocol in question. The text of any proposed amendment to this Convention or to any protocol except as may otherwise be provided in such protocol shall be communicated to the Parties to the instrument in question by the secretariat at least six months before the meeting at which it is proposed for adoption. The secretariat shall also communicate proposed amendments to the signatories to this Convention for information.

3. The Parties shall make every effort to reach agreement on any proposed amendment to this Convention or to any protocol by consensus. If all efforts at consensus have been exhausted and no agreement reached the amendment shall as a last resort be adopted by a two-third majority vote of the Parties to the instrument in question present and voting at the meeting and shall be submitted by the Depositary to all Parties for ratification acceptance or approval.

4. Ratification acceptance or approval of amendments shall be notified to the Depositary in writing. Amendments adopted in accordance with paragraph 3 above shall enter into force among Parties having accepted them on the ninetieth day after the deposit of instruments of ratification acceptance or approval by at least two thirds of the Contracting Parties to this Convention or of the Parties to the protocol concerned except as may otherwise be provided in such protocol. Thereafter the amendments shall enter into force for any other Party on the ninetieth day after that Party deposits its instrument of ratification acceptance or approval of the amendments.

5. For the purpose of this Article. Parties present and voting means Parties present and casting an affirmative or negative vote.

Article 30

Adoption and Amendment of Annexes

1. The annexes to this Convention or to any protocol shall form an integral part of the Convention or of such protocol, as the case may be, and, unless expressly provided otherwise, a reference to this Convention or its protocols constitutes at the same time a reference to any annexes thereto. Such annexes shall be restricted to procedural, scientific, technical and administrative matters.

2. Except as may be otherwise provided in any protocol with respect to its annexes, the following procedure shall apply to the proposal, adoption and entry into force of additional annexes to this Convention or of annexes to any protocol:

(a) Annexes to this Convention or to any protocol shall be proposed and adopted according to the procedure laid down in Article 29;

(b) Any Party that is unable to approve an additional annex to this Convention or an annex to any protocol to which it is Party shall so notify the Depositary, in writing within one year from the date of the communication of the adoption by the Depositary. The Depositary shall without delay notify all Parties of any such notification received. A Party may at any time withdraw a previous declaration of objection and the annexes shall thereupon enter into force for that Party subject to subparagraph (c) below;

(c) On the expiry of one year from the date of the communication of the adoption by the Depositary, the annex shall enter into force for all Parties to this Convention or to any protocol concerned which have not submitted a notification in accordance with the provisions of subparagraph (b) above.

3. The proposal adoption and entry into force of amendments to annexes to this Convention or to any protocol shall be subject to the same procedure as for the proposal adoption and entry into force of annexes to the Convention or annexes to any protocol.

4. If an additional annex or an amendment to an annex is related to an amendment to this Convention or to any protocol, the additional annex or amendment shall not enter into force until such time as the amendment to the Convention or to the protocol concerned enters into force.

Article 31

Right to Vote

1. Except as provided for in paragraph 2 below, each Contracting Party to this Convention or to any protocol shall have one vote.

2. Regional economic integration organizations, in matters within their competence, shall exercise their right to vote with a number of votes equal to the number of their member States which are Contracting

Parties to this Convention or the relevant protocol. Such organizations shall not exercise their right to vote if their member States exercise theirs, and *vice versa*.

Article 32

Relationship between this Convention and its Protocols

1. A State or a regional economic integration organization may not become a Party to a protocol unless it is, or becomes at the same time, a Contracting Party of this Convention.

2. Decisions under any protocol shall be taken only by the Parties to the protocol concerned. Any Contracting Party that has not ratified, accepted or approved a protocol may participate as an observer in any meeting of the parties to the protocol.

Article 33

Signature

This Convention shall be open for signature at Rio de Janeiro by all States and any regional economic integration organization from 5 June 1992 until 14 June 1992, and at the United Nations Headquarters in New York from 15 June 1992 to 4 June 1993.

Article 34

Ratification, Acceptance or Approval

1. The Convention and any protocol shall be subject to ratification acceptance or approval by States and by regional economic integration organizations. Instruments of ratification, acceptance or approval shall be deposited with the Depositary.

2. Any organization referred to in paragraph 1 above which becomes a Contracting Party to this Convention or any protocol without any of its

member States being a Contracting Party shall be bound by all the obligations under the Convention or the protocol, as the case may be. In the case of such organizations, one or more of whose member States is a Contracting Party to this Convention or relevant protocol, the organization and its member States shall decide on their respective responsibilities for the performance of their obligations under the Convention or protocol, as the case may be. In such cases, the organization and the member States shall not be entitled to exercise rights under the Convention or relevant protocol concurrently.

3. In their instruments of ratification, acceptance or approval, the organizations referred to in paragraph 1 above shall declare the extent of their competence with respect to the matters governed by the Convention or the relevant protocol. These organizations shall also inform the Depositary of any relevant modification in the extent of their competence.

Article 35

Accession

1. This Convention and any protocol shall be open for accession by States and by regional economic integration organizations from the date on which the Convention or the protocol concerned is closed for signature. The instruments of accession shall be deposited with the Depositary.

2. In their instruments of accession the organizations referred to in paragraph 1 above shall declare the extent of their competence with respect to the matters governed by the Convention or the relevant protocol. These organizations shall also inform the Depositary of any relevant modification in the extent of their competence.

3. The provisions of Article 34, paragraph 2, shall apply to regional economic integration organizations which accede to this Convention or any protocol.

Article 36

Entry into Force

1. This Convention shall enter into force on the ninetieth day after the date of deposit of the thirtieth instrument of ratification, acceptance, approval or accession.

2. Any protocol shall enter into force on the ninetieth day after the date of deposit of the number of instruments of ratification, acceptance, approval or accession, specified in that protocol, has been deposited.

3. For each Contracting Party, which ratifies, accepts or approves this Convention or accedes thereto after the deposit of the thirtieth instrument of ratification, acceptance, approval or accession, it shall enter into force on the ninetieth day after the date of deposit by such Contracting Party of its instrument of ratification, acceptance, approval or accession.

4. Any protocol, except as otherwise provided in such protocol, shall enter into force for a Contracting Party that ratifies, accepts or approves that protocol or accedes thereto after its entry into force pursuant to paragraph 2 above, on the ninetieth day after the date on which that Contracting Party deposits its instrument of ratification, acceptance, approval or accession, or on the date on which this Convention enters into force for that Contracting Party whichever shall be the later.

5. For the purposes of paragraphs 1 and 2 above any instrument deposited by a regional economic integration organization shall not be counted as additional to those deposited by member States of such organization.

Article 37

Reservations

No reservations may be made to this Convention.

Article 38

Withdrawls

1. At any time after two years from the date on which this Convention has entered into force for a Contracting Party, that Contracting Party may withdraw from the Convention by giving written notification to the Depositary.
2. Any such withdrawal shall take place upon expiry of one year after the date of its receipt by the Depositary, or on such later date as may be specified in the notification of the withdrawal.

3. Any Contracting Party which withdraws from this convention shall be considered as also having withdrawn from any protocol to which it is party.

Article 39

Financial Interim Arrangements

Provided that it has been fully restructured in accordance with the requirements of Article 21, the Global Environment Facility of the United Nations Development Programme, the United Nations Environment Programme and the International Bank for Reconstruction and Development shall be the institutional structure referred to in Article 21 on an interim basis, for the period between the entry into force of this Convention and the first meeting of the Conference of the Parties or until the Conference of the Parties decides which institutional structure will be designated in accordance with Article 21.

Article 40

Secretariat Interim Arrangements

The secretariat to be provided by the Executive Director of the United Nations Environment Programme shall be the secretariat referred to in Article 24, paragraph 2, on an interim basis for the period between the entry into force of this Convention and the first meeting of the Conference of the Parties.

Article 41

Depositary

The Secretary General of the United Nations shall assume the functions of Depositary of this Convention and any protocols.

Article 42

Authentic Texts

The original of this Convention, of which the Arabic, Chinese, English, French, Russian and Spanish texts are equally authentic, shall be deposited with the secretary-general of the United Nations.

IN WITNESS WHERE OF the undersigned, being duly authorized to that effect, have signed this Convention.

Done at Rio de Janeiro on this fifth day of June, one thousand nine hundred and ninety-two.

ANNEXURE I

Identification and Monitoring

1. Ecosystems and habitats: containing high diversity, large numbers of endemic or threatened species, or wilderness; required by migratory species; of social, economic, cultural or scientific importance; or, which are representative, unique or associated with key evolutionary or other biological processes;

2. Species and communities which are: threatened; wild relatives of domesticated or cultivated species; of medicinal, agricultural or other economic value; or social, scientific or cultural importance; or importance for research into the conservation and sustainable use of biological diversity, such as indicator species; and

3. Described genomes and genes of social, scientific or economic importance.

ANNEXURE II

Part I

Arbitration

Article 1

The claimant party shall notify the secretariat that the parties are referring a dispute to arbitration pursuant to Article 27. The notification shall state the subject-matter of arbitration and include, in particular, the articles of the Convention or the protocol, the interpretation or application of which are at issue. If the parties do not agree on the subject matter of the dispute before the President of the tribunal is designated, the arbitrate tribunal shall determine the subject matte. The secretariat shall forward the information thus received to all Contracting Parties to this Convention or to the protocol concerned.

Article 2

1. In disputes between two parties, the arbitrate tribunal shall consist of three members. Each of the parties to the dispute shall appoint an arbitrator and the two arbitrators so appointed shall designate by common agreement the third arbitrator who shall be the President of the tribunal. The latter shall not be a national of one of the parties to the dispute, nor have his or her usual place of residence in the territory of one of these parties, nor be employed by any of them, nor have dealt with the case in any other capacity.

2. In disputes between more than two parties, parties in the same interest shall appoint one arbitrator jointly by agreement.

3. Any vacancy shall be filled in the manner prescribed for the initial appointment.

Article 3

1. If the President of the arbitrate tribunal has not been designated within two months of the appointment of the second arbitrator, the Secretary

General of the United Nations shall, at the request of a party, designate the President within a further two-month period.

2. If one of the parties to the dispute does not appoint an arbitrator within two months of receipt of the request, the other party may inform the secretary-general who shall make the designation within a further two-month period.

Article 4

The arbitrate tribunal shall render its decisions in accordance with the provisions of this Convention, any protocols concerned and international law.

Article 5

Unless the parties to the dispute otherwise agree, the arbitrate tribunal shall determine its own rules of procedure.

Article 6

The arbitrate tribunal may, at the request of one of the parties, recommend essential interim measures of protection.

Article 7

The parties to the dispute shall facilitate the work of the arbitrate tribunal and, in particular, using all means at their disposal, shall:

(a) Provide it with all relevant documents, information and facilities; and

(b) Enable it when necessary, to call witnesses or experts and receive their evidence.

Article 8

The parties and the arbitrators are under an obligation to protect the confidentiality of any information they receive in confidence during the proceedings of the arbitrate tribunal.

Article 9

Unless the arbitrate tribunal determines otherwise because of the particular circumstances of the case, the costs of the tribunal shall be borne by the parties to the dispute in equal shares. The tribal shall keep a record of all its costs, and shall furnish a final statement thereof to the parties.

Article 10

Any Contracting Party that has an interest of a legal nature in the subject-matter of the dispute which may be affected by the decision in the case, may intervene in the proceedings with the consent of the tribunal.

Article 11

The tribunal may hear and determine counterclaims arising directly out of the subject-matter of the dispute.

Article 12

Decisions both on procedure and substance of the arbitrate tribunal shall be taken by a majority vote of its members.

Article 13

If one of the parties to the dispute does not appear before the arbitrate tribunal or fails to defend its case, the other party may request The tribunal to continue the proceedings and to make its award. Absence of a party or a failure of a party to defend its case shall not constitute a bar to the proceedings. Before rendering its final decision, the arbitrate tribunal must satisfy itself that the claim is well founded in fact and law.

Article 14

The tribunal shall render its final decision within five months of the date on which it is fully constituted unless it finds it necessary to extend the time limit for a period which should not exceed five more months.

Article 15

The final decision of the arbitrate tribunal shall be confined to the subject matter of the dispute and shall state the reasons on which it is based. It shall contain the names of the members who have participated and the date of the final decision. Any member of the tribunal may attach a separate or dissenting opinion to the final decision.

Article 16

The award shall be binding on the parties to the dispute. It shall be without appeal unless the parties to the dispute have agreed in advance to an appellate procedure.

Article 17

Any controversy which may arise between the parties to the dispute as regards the interpretation or manner of implementation of the final decision may be submitted by either party for decision to the arbitrate tribunal which rendered it.

THE BIOLOGICAL DIVERSITY ACT, 2002

Ministry of Law and Justice (Legislative Department) New Delhi, the 5th February, 2003

The Biological Diversity Act, 2002 / No. 18 of 2003 [5th February, 2003]

Sr. No. Description of the clause

1 The Biological Diversity Act

CHAPTER I

PRELIMINARY

Sr. No. Description of the clause

1 Short title, extent and commencement

2 Definitions

CHAPTER II

Regulation of Access To Biological Diversity

Sr. No. Description of the clause

3 Certain persons not to undertake Biodiversity related activities without approval of National Biodiversity Authority

4 Results of research not to be transferred to certain persons without approval of National Biodiversity Authority

5 Sections 3 and 4 not to apply to certain collaborative research projects

6 Application for intellectual- property rights not to be made without approval of National Biodiversity Authority

7 Prior intimation to State Biodiversity Board for obtaining biological resource for certain purposes

CHAPTER III

National Biodiversity Authority

Sr. No. Description of the clause

8 Establishment of National Biodiversity Authority

9 Conditions of service of Chairperson and Members

10 Chairperson to be Chief Executive of National Biodiversity Authority

11 Removal of members

12 Meetings of National Biodiversity Authority

13 Committees of National Biodiversity Authority

14 Officers and employees of National Biodiversity Authority

15 Authentication of orders and decisions of National Biodiversity Authority

16 Delegation of powers

17 Expenses of National Biodiversity Authority to be defrayed out of the Consolidated Fund of India

CHAPTER IV

Functions and powers of the National Biodiversity Authority

Sr. No. Description of the clause

18 Functions and powers of National Biodiversity Authority

CHAPTER V

Approval by the National Biodiversity Authority

Sr. No.	Description of the clause
19	Approval by National Biodiversity Authority for undertaking certain activities
20	Transfer of biological resource or knowledge
21	Determination of equitable benefit sharing by National Biodiversity Authority

CHAPTER VI

State Biodiversity Board

Sr. No.	Description of the clause
22	Establishment of State Biodiversity Board
23	Functions of State Biodiversity Board
24	Power of State Biodiversity Board to restrict certain activities violating the objectives of conservation, etc.
25	Provisions of sections 9 to 17 to apply with modifications to State Biodiversity Board

CHAPTER VII

Finance, accounts and audit of National Biodiversity Authority

Sr. No.	Description of the clause
26	Grants or loans by the Central Government

27 Constitution of National Biodiversity Fund

28 Annual report of National Biodiversity Authority

29 Budget, accounts and audit

30 Annual report to be laid before Parliament

CHAPTER VIII

Finance, Accounts And Audit of State Biodiversity Board

Sr. No. Description of the clause

31 Grant of money by State Government to State Biodiversity Board

32 Constitution of State Biodiversity Fund

33 Annual report of State Biodiversity Board

34 Audit of accounts of State Biodiversity Board

35 Annual report of State Biodiversity Board to be laid before State Legislature

CHAPTER IX

Duties of the Central and the State Governments

Sr. No. Description of the clause

36 Central Government to develop National strategies, plans, etc., for conservation, etc., of biological diversity

37 Biodiversity heritage sites

38 Power of Central Government to notify threatened species

CHAPTER X

Biodiversity Management Committees

Sr. No. Description of the clause

CHAPTER XI

Local Biodiversity Fund

Sr. No. Description of the clause

CHAPTER XII

MISCELLANEOUS

Sr. No. Description of the clause

Ministry of Law and Justice

(Legislative Department)
New Delhi , the 5th February, 2003/Magha 16, 1924 (Saka)

The following Act of Parliament received the assent of the President on the 5th February, 2003 , and is hereby published for general information:

THE BIOLOGICAL DIVERSITY ACT, 2002

No. 18 OF 2003

[5th February, 2003]

An Act to provide for conservation of biological diversity, sustainable use of its components and fair and equitable sharing of the benefits arising out of the use of biological resources, knowledge and for matters connected therewith or incidental thereto.

WHEREAS India is rich in biological diversity and associated traditional and contemporary knowledge system relating thereto.

AND WHEREAS India is a party to the United Nations Convention on Biological Diversity signed at Rio de Janeiro on the 5th day of June, 1992;

AND WHEREAS the said Convention came into force on the 29th December, 1993 ;

AND WHEREAS the said Convention reaffirms the sovereign rights of the States over their biological resources;

AND WHEREAS the said Convention has the main objective of conservation of biological diversity, sustainable use of its components and fair and equitable sharing of the benefits arising out of utilization of genetic resources;

AND WHEREAS it is considered necessary to provide for conservation, sustainable utilization and equitable sharing of the benefits arising out of utilization of genetic resources and also to give effect to the said Convention.

BE it enacted by Parliament in the Fifty-third Year of the Republic of India as follows:–

CHAPTER - I

Preliminary

1. Short title, extent and com-mencement

(1) This Act may be called the Biological Diversity Act, 2002.

(2) It extends to the whole of India .

(3) It shall come into force on such date as the Central Government may, by notification in the Official Gazette, appoint:

Provided that different dates may be appointed for different provisions of this Act and any reference in any such provision to the commencement of this Act shall be construed as a reference to the coming into force of that provision.

2. Definitions

In this Act, unless the context otherwise requires,–

(a) "benefit claimers" means the conservers of biological resources, their byproducts, creators and holders of knowledge and information relating to the use of such biological resources, innovations and practices associated with such use and application;

(b) "biological diversity" means the variability among living organisms from all sources and the ecological complexes of which they are part, and includes diversity within species or between species and of eco-systems;

(c) "biological resources" means plants, animals and micro-organisms or parts thereof, their genetic material and by-products (excluding value added products) with actual or potential use or value, but does not include human genetic material;

(d) "bio-survey and bio-utilization" means survey or collection of species, subspecies, genes, components and extracts of biological resource for any purpose and includes characterization, inventorisation and bioassay;

(e) "Chairperson" means the Chairperson of the National Biodiversity Authority or, as the case may be, of the State Biodiversity Board;

(f) "commercial utilization" means end uses of biological resources for commercial utilization such as drugs, industrial enzymes, food flavours, fragrance, cosmetics, emulsifiers, oleoresins, colours, extracts and genes used for improving crops and livestock through genetic intervention, but does not include conventional breeding or traditional practices in use in any agriculture, horticulture, poultry, dairy farming, animal husbandry or bee keeping;

(g) "fair and equitable benefit sharing" means sharing of benefits as determined by the National Biodiversity Authority under section 21;

(h) "local bodies" means Panchayats and Municipalities, by whatever name called, within the meaning of clause (1) of article 243B and clause (1) of article 243Q of the Constitution and in the absence of any Panchayats or Municipalities, institutions of self-government constituted under any other provision of the Constitution or any Central Act or State Act;

(i) "member" means a member of the National Biodiversity Authority or a State Biodiversity Board and includes the Chairperson;

(j) "National Biodiversity Authority" means the National Biodiversity Authority established under section 8;

(k) "prescribed" means prescribed by rules made 1 under this Act;

(l) "regulations" means regulations made under this Act;

(m) "research" means study or systematic investigation of any biological resource or technological application, that uses biological systems, living organisms or derivatives thereof to make or modify products or processes for any use;

(n) "State Biodiversity Board" means the State Biodiversity Board established under section 22;

(o) "sustainable use" means the use of components of biological diversity in such manner and at such rate that does not lead to the long-term decline of the biological diversity thereby maintaining its potential to meet the needs and aspirations of present and future generations;

(p) "value added products" means products which may contain portions or extracts of plants and animals in unrecognizable and physically inseparable form.

CHAPTER - II

Regulation of Access to Biological Diversity

Certain persons not to undertake Biodiversity related activities without approval of National Biodiversity Authority

3. (1) No person referred to in sub-section (2) shall, without previous approval of the National Biodiversity Authority, obtain any biological resource occurring in India or knowledge associated thereto for research or for commercial utilization or for bio-survey and bio-utilization.

(2) The persons who shall be required to take the approval of the National Biodiversity Authority under sub-section (1) are the following, namely:

(a) a person who is not a citizen of India ;

(b) a citizen of India , who is a non-resident as defined in clause (30) of section 2 of the Income-tax Act, 1961;

(c) a body corporate, association or organization-

(i) not incorporated or registered in India ; or

(ii) incorporated or registered in India under any law for the time being in force which has any non-Indian participation in its share capital or management.

Results of research not to be transferred to certain persons without approval of National Biodiversity Authority

4. No person shall, without the previous approval of the National Biodiversity Authority, transfer the results of any research relating to any biological resources occurring in, or obtained from, India for monetary consideration or otherwise to any person who is not a citizen of India or citizen of India who is non-resident as defined in clause (30) of section 2 of the Income-tax Act, 1961 or a body corporate or organization which Is not registered or incorpo-rated in India or which has any non-Indian participation in its share capital or management.

Explanation.- For the purposes of this section, "transfer" does not include publication of research papers or dissemination of knowledge in any seminar or workshop, if such publication is as per the guidelines issued by the Central Government.

Sections 3 and 4 not to apply to certain collaborative research projects

5.(1) The provisions of sections 3 and 4 shall not apply to collaborative research projects involving transfer or exchange of biological resources or information relating thereto between institutions, including Government sponsored institutions of India, and such institutions in other countries, if such collaborative research projects satisfy the conditions specified in sub-section (3).

(2) All collaborative research projects, other than those referred to in sub-section (1) which are based on agreements concluded before the commencement of this Act and in force shall, to the extent the provisions of agreement are inconsistent with the provisions of this Act or any guidelines issued under clause (a) of sub-section (3), be void.

(3) For the purposes of sub-section (1), collaborative research projects shall-

(a) conform to the policy guidelines issued by the Central Government in this behalf;

(b) be approved by the Central Government.

Application for intellectual property rights not to be made without approval of National Biodiversity Authority.

6.(1) No person shall apply for any intellectual property right, by whatever name called, in or outside India for any invention based on any research or information on a biological resource obtained from India without obtaining the previous approval of the National Biodiversity Authority before making such application.

Provided that if a person applies for a patent, permission of the National Biodiversity Authority may be obtained after the acceptance of the patent but before the seating of tile patent by the patent authority concerned:

Provided further that the National Biodiversity Authority shall dispose of the application for permission made to it within a period of ninety days from the date of receipt thereof.

(2) The National Biodiversity Authority may, while granting the approval under this section, impose benefit sharing fee or royalty or both or impose conditions including the sharing of financial benefits arising out of the commercial utilization of such rights.

(3) The provisions of this section shall not apply to any person making an application for any right under any law relating to protection of plant varieties enacted by Parliament.

(4) Where any right is granted under law referred to in sub-section (3), the concerned authority granting such right shall endorse a copy of such document granting the right to the National Biodiversity Authority.

Prior intimation to State Biodiversity Board for obtaining biological resource for certain purposes

7. No person, who is a citizen of India or a body corporate, association or organization which is registered in India , shall obtain any biological resource for commercial utilization, or bio-survey and bio-utilization for commercial utilization except after giving prior intimation to the State Biodiversity Board concerned:

Provided that the provisions of this section shall not apply to the local people and communities of the area, including growers and cultivators of biodiversity, and vaids and hakims, who have been practicing indigenous medicine.

CHAPTER - III

National Biodiversity Authority

Establishment of National Biodiversity Authority

8.(1) With effect from such date as the Central Government may, by notification in the Official Gazette, appoint, there shall be established by the Central Government for the purposes of this Act, a body to be called the National Biodiversity Authority.

(2) The National Biodiversity Authority shall be a body corporate by the name aforesaid, having perpetual succession and a common seal, with power to acquire, hold and dispose of property, both movable and immovable, and to contract, and shall by the said name sue and be sued.

(3) The head office of the National Biodiversity Authority shall be at Chennai and the National Biodiversity Authority may, with the previous approval of the Central Government, establish offices at other places in India .

(4) The National Biodiversity Authority shall consist of the following members,

(a) a Chairperson, who shall be an eminent person having adequate knowledge and experience in the conservation and sustainable use of biological diversity and in matters relating to equitable sharing of benefits, to be appointed by the Central Government;

(b) three ex officio members to be appointed by the Central Government, one representing the Ministry dealing with Tribal Affairs and two representing the Ministry dealing with Environment and Forests of whom one shall be the Additional Director General of Forests or the Director General of Forests;

(c) seven ex officio members to be appointed by the Central Government to represent respectively the Ministries of the Central Government dealing with –

(i) Agricultural Research and Education;

(ii) Biotechnology;

(iii) Ocean Development;

(iv) Agriculture and Cooperation;

(v) Indian Systems of Medicine and Homoeopathy;

(vi) Science and Technology;

(vii) Scientific and Industrial Research;

(d) five non-official members to be appointed from amongst specialists and scientists having special knowledge of, or experience in, matters relating to conservation of biological diversity, sustainable use of biological resources and equitable sharing of benefits arising out of the use of biological resources, representatives of industry, conservers, creators and know ledge-holders of biological resources.

Conditions of service of Chairperson and members

9. The term of office and conditions of service of the Chairperson and the other members other than ex officio members of the National Biodiversity Authority shall be such as may be prescribed by the Central Government.

Chairperson to be Chief Executive of National Biodiversity Authority

10. The Chairperson shall be the Chief Executive of the National Biodiversity Authority and shall exercise such powers and perform such duties, as may be prescribed.

Removal of members

11. The Central Government may remove from the National Biodiversity Authority any member who, in its opinion, has-

(a) been adjudged as an insolvent; or

(b) been convicted of an offence which involves moral turpitude; or

(c) become physically or mentally incapable of acting as a member; or

(d) so abused his position as to render his continuance in office detrimental to the public interest; or

(e) acquired such financial or other interest as is likely to affect prejudicially his functions as a member.

Meetings of National Biodiversity Authority

12. (1) The National Biodiversity Authority shall meet at such time and place and shall observe such rules of procedure in regard to the transaction of business at its meetings (including the quorum at its meetings) as may be prescribed.

(2) The Chairperson of the National Biodiversity Authority shall preside at the meetings of the National Biodiversity Authority.

(3) If for any reason the Chairperson is unable to attend any meeting of the National Biodiversity Authority, any member of the National Biodiversity Authority chosen by the members present at the meeting shall preside at the meeting.

(4) All questions which come before any meeting of the National Biodiversity Authority shall be decided by a majority of votes of the members present and voting and in the event of equality of votes, the Chairperson or, in his absence, the person presiding, shall have and exercise a second or casting vote.

(5) Every member who is in any way, whether directly, indirectly or personally, concerned or interested in a matter to be decided at the meeting

shall disclose the nature of his concern or interest and after such disclosure, the member concerned or interested shall not attend that meeting.

(6) No act or proceeding of the National Biodiversity Authority shall be invalidated merely by reason of-

(a) any vacancy in, or any defect in the constitution of, the National Biodiversity Authority; or

(b) any defect in the appointment of a person acting as a member; or

(c) any irregularity in the procedure of the National Biodiversity Authority not affecting the merits of the case.

Committees of National Biodiversity Authority

13. (1) The National Biodiversity Authority may constitute a committee to deal with agro-biodiversity.

Explanation- For the purposes of this sub-section, "agro-biodiversity" means bio-logical diversity, of agriculture related species and their wild relatives.

(2) Without prejudice to the provisions of sub-section (1), the National Biodiversity Authority may constitute such number of committees as it deems fit for the efficient dis-charge of its duties and performance of its functions under this Act.

(3) A committee constituted under this section shall co-opt such number of persons, who are not the members of the National Biodiversity Authority, as it may think fit and the persons so co-opted shall have the right to attend the meetings of the committee and take part in its proceedings but shall not have the right to vote.

(4) The persons appointed as members of the committee under sub-section (2) shall be entitled to receive such allowances or fees for attending the meetings of the committee as may be fixed by the Central Government.

Officers and employees of National Biodiversity Authority

14.(1) The National Biodiversity Authority may appoint such officers and other employees as it considers necessary for the efficient discharge of its functions under this Act.

(2) The terms and conditions of service of such officers and other employees of the National Biodiversity Authority shall be such as may be specified by regulations.

Authentication of orders and decisions of National Biodiversity Authority

15. All orders and decisions of the National Biodiversity Authority shall be authenticated by the signature of the Chairperson or any other member authorized by the National Biodiversity Authority in this behalf and all other instruments executed by the National Biodiversity Authority shall be authenticated by the signature of an officer of the National Biodiversity Authority authorized by it in this behalf.

Delegation of powers

16. The National Biodiversity Authority may, by general or special order in writing, delegate to any member, officer of the National Biodiversity Authority or any other person subject to such conditions, if any, as may be specified in the order, such of the powers and functions under this Act (except the power to prefer an appeal under section 50 and the power to make regulations under section 64) as it may deem necessary.

Expenses of National Biodiversity Authority to be defrayed out of the Consolidated Fund of India

17. The salaries and allowances payable to the members and the administrative expenses of the National Biodiversity Authority including salaries, allowances and pension payable to, or in respect of, the officers and other employees of the National Biodiversity Authority shall be defrayed out of the Consolidated Fund of India.

CHAPTER - IV

Functions and Powers of the National Biodiversity Authority

18.(1) It shall be the duty of the National Biodiversity Authority to regulate activities referred to in sections 3, 4 and 6 and by regulations issue guidelines for access to biological resources and for fair and equitable benefit sharing.

(2) The National Biodiversity Authority may grant approval for undertaking any activity referred to in sections 3, 4 and 6.

(3) The National Biodiversity Authority may-

(a) advise the Central Government on matters relating to the conservation of biodiversity, sustainable use of its components and equitable sharing of benefits arising out of the utilization of biological resources;

(b) advise the State Governments in the selection of areas of biodiversity importance to be notified under sub-section (1) of section 37 as heritage sites and measures for the management of such heritage sites;

(c) perform such other functions as may be necessary to carry out the provisions of this Act.

(4) The National Biodiversity Authority may, on behalf of the Central Government, take any measures necessary to oppose the grant of intellectual property rights in any country outside India on any biological resource obtained from India or knowledge associated with such biological resource which is derived from India .

CHAPTER - V

Approval by the National Biodiversity Authority

Approval by National Biodiversity Authority for undertaking certain activities

19.(1) Any person referred. to in sub-section (2) of section 3 who intends to obtain any biological resource occurring in India or knowledge associated thereto for research or for commercial utilization or for bio-survey and bio-utilization or transfer the results of any research relating to biological resources occurring in, or obtained from, India, shall make application in such form and payment of such fees as may be prescribed, to the National Biodiversity Authority.

(2) Any person who intends to apply for a patent or any other form of intellectual property protection whether in India or outside India referred to in sub-section (1) of section 6, may make an application in such form and in such manner as may be prescribed to the National Biodiversity Authority.

(3) On receipt of an application under sub-section (1) or sub-section (2), the National Biodiversity Authority may, after making such enquiries as it may deem fit and if necessary after consulting an expert committee constituted for this purpose, by order, grant approval subject to any regulations made in this behalf and subject to such terms and conditions as it may deem fit, including the imposition of charges by way of royalty or for reasons to be recorded in writing, reject the application:

Provided that no such order for rejection shall be made without giving an opportunity of being heard to the person affected.

(4) The National Biodiversity Authority shall give public notice of every approval granted by it under this section.

Transfer of biological resource or knowledge

20.(1) No person who has been granted approval under section 19 shall transfer any biological resource or knowledge associated thereto which is the subject matter of the said approval except with the permission of the National Biodiversity Authority.

(2) Any person who intends to transfer any biological resource or knowledge associated thereto referred to in sub-section (1) shall make an application in such form and in such manner as may be prescribed to the National Biodiversity Authority.

(3) On receipt of an application under sub-section (2), the National Biodiversity Authority may, after making such enquiries as it may deem fit and if necessary after consulting an expert committee constituted for this

purpose, by order, grant approval subject to such terms and conditions as it may deem fit, including the imposition of charges by way of royalty or for reasons to be recorded in writing, reject the application:

Provided that no such order for rejection shall be made without giving an opportunity of being heard to the person affected.

(4) The National Biodiversity Authority shall give public notice of every approval granted by it under this section.

Determination of equitable benefit sharing by National Biodiversity Authority

21.(1) The National Biodiversity Authority shall while granting approvals under section 19 or section 20 ensure that the terms and conditions subject to which approval is granted secures equitable sharing of benefits arising out of the use of accessed biological resources, their by-products, innovations and practices associated with their use and applications and knowledge relating thereto in accordance with mutually agreed terms and conditions between the person applying for such approval, local bodies concerned and the benefit claimers.

(2) The National Biodiversity Authority shall, subject to any regulations made in this behalf, determine the benefit sharing which shall be given effect in all or any of the following manner, namely:-

(a) grant of joint ownership of intellectual property rights to the National Biodiversity Authority, or where benefit claimers are identified, to such benefit claimers;

(b) transfer of technology;

(c) location of production, research and development units in such areas which will facilitate better living standards to the benefit claimers;

(d) association of Indian scientists, benefit claimers and the local people with research and development in biological resources and bio-survey and bio-utilization;

(e) setting up of venture capital fund for aiding the cause of benefit claimers;

(f) payment of monetary compensation and non-monetary benefits to the benefit claimers as the National Biodiversity Authority may deem fit.

(3) Where any amount of money is ordered by way of benefit sharing, the National Biodiversity Authority may direct the amount to be deposited in the National Biodiversity Fund:

Provided that where biological resource or knowledge was a result of access from specific individual or group of individuals or organizations, the National Biodiversity Authority may direct that the amount shall be paid directly to such individual or group of individuals or organizations in accordance with the terms of any agreement and in such manner as it deems fit.

(4) For the purposes of this section, the National Biodiversity Authority shall, in consultation with the Central Government, by regulations, frame guidelines.

CHAPTER - VI

State Biodiversity Board

Establishment of State Biodiversity Board

22.(1) With effect from such date as the State Government may, by notification in the Official Gazette, appoint in this behalf, there shall be established by that Government for the purposes of this Act, a Board for the State to be known as the (name of the State) Biodiversity Board.

(2) Notwithstanding anything contained in this section, no State Biodiversity Board shall be constituted for a Union territory and in relation to a Union territory, the National Biodiversity Authority shall exercise the powers and perform the functions of a State Biodiversity Board for that Union territory:

Provided that in relation to any Union territory, the National Biodiversity Authority may delegate all or any of its powers or functions under this sub-section to such person or group of persons as the Central Government may specify.

(3) The Board shall be a body corporate by the name aforesaid, having perpetual succession and a common seal, with power to acquire, hold and dispose of property, both movable and immovable, and to contract, and shall by the said name sue and he sued.

(4) The Board shall consist of the following members, namely:

(a) a Chairperson who shall be an eminent person having adequate knowledge and experience in the conservation and sustainable use of biological diversity and in matters relating to equitable sharing of benefits, to he appointed by the State Government;

(b) not more than five ex officio members to be appointed by the State Government to represent the concerned Departments of the State Government;

(c) not more than five members to be appointed from amongst experts in matters relating to conservation of biological diversity, sustainable use of biological resources and equitable sharing of benefits arising out of the use of biological resources.

(5) The head office of the State Biodiversity Board shall be at such place as the State Government may, by notification in the Official Gazette, specify.

Functions of State Biodiversity Board

23. The functions of the State Biodiversity Board shall be to–

(a) advise the State Government, subject to any guidelines issued by the Central Government, on matters relating to the conservation of biodiversity, sustainable use of its components and equitable sharing of the benefits arising out of the utilization of biological resources;

(b) regulate by granting of approvals or otherwise requests for commercial utilization or bio-survey and bio-utilization of any biological resource by Indians;

(c) perform such other functions as may he necessary to carry out the provisions of this Act or as may be prescribed by the State Government.

Power of State Biodiversity Board to restrict certain activities

24.(1) Any citizen of India or a body corporate, organization or association registered in India intending to undertake any activity referred to in section 7 shall give prior intimation in such form as may be prescribed by the State Government to the State Biodiversity Board.

(2) On receipt of an intimation under sub-section (1), the State Biodiversity Board may, in consultation with the local bodies concerned and after making such enquires as it conservation, may deem fit, by order, prohibit or restrict any such activity if it is of opinion that such activity is detrimental or contrary to the objectives of conservation and sustainable use of biodiversity or equitable sharing of benefits arising out of such activity:

Provided that no such order shall be made without giving an opportunity of being heard to the person affected.

(3) Any information given in the form referred to in sub-section (1) for prior intimation shall be kept confidential and shall not be disclosed, either intentionally or unintentionally, to any person not concerned thereto.

Provisions of sections 9 to 17 to apply with modifications to State Biodiversity Board

25. The provisions of sections 9 to 17 shall apply to a State Biodiversity Board and shall have effect subject to the following modifications, namely:

(a) references to the Central Government shall be construed as references to the State Government;

(b) references to the National Biodiversity Authority shall be construed as references to the State Biodiversity Board;

Provisions of sections 9 to 17 to apply with modifications to State Biodiversity Board.

(c) reference to the Consolidated Fund of India shall be construed as reference to the Consolidated Fund of the State.

CHAPTER - VII

Finance, Accounts and Audit of National Biodiversity Authority

Grants or loans by the Central Government

26. The Central Government may, after due appropriation made by Parliament by law in this behalf, pay to the National Biodiversity Authority by way of grants or loans such sums of money as the Central Government may think fit for being utilized for the purposes of this Act.

Constitution of National Biodiversity Fund

27.(1) There shall be constituted a Fund to be called the National Biodiversity Fund and there shall be credited thereto --

(a) any grants and loans made to the National Biodiversity Authority under section 26;

(b) all charges and royalties received by the National Biodiversity Authority under this Act; and

(c) all sums received by the National Biodiversity Authority from such other sources as may be decided upon by the Central Government.

(2) The Fund shall be applied for—

(a) channeling benefits to the benefit claimers;

(b) conservation and promotion of biological resources and development of areas from where such biological resources or knowledge associated thereto has been accessed;

(c) socio-economic development of areas referred to in clause (b) in consultation with the local bodies concerned.

Annual report of National Biodiversity Authority

28. The National Biodiversity Authority shall prepare, in such form and at such time each financial year as may be prescribed, its annual report, giving a full account of its activities during the previous financial year and furnish, to the Central Government, before such date as may be prescribed, its audited copy of accounts together with auditors report thereon.

Budget accounts and audit

29.(1) The National Biodiversity Authority shall prepare a budget, maintain proper accounts and other relevant records (including the accounts and other relevant records of the National Biodiversity Fund) and prepare an annual statement of account in such form as may be prescribed by the Central Government in consultation with the Comptroller and Auditor-General of India.

(2) The accounts of the National Biodiversity Authority shall be audited by the Comptroller and Auditor-General of India at such intervals as may be specified by him and any expenditure incurred in connection with such audit shall be payable by the National Biodiversity Authority to the Comptroller and Auditor-General of India.

(3) The Comptroller and Auditor-General of India and any other person appointed by him in connection with the audit of the accounts of the National Biodiversity Authority shall have the same rights and privileges and authority in connection with such audit as the Comptroller and Auditor-General generally has in connection with the audit of the Government accounts and, in particular, shall have the right to demand the production of books, accounts, connected vouchers and other documents and papers and to inspect any of the offices of the National Biodiversity Authority.

(4) The accounts of the National Biodiversity Authority as certified by the Comptroller and Auditor- General of India or any other person appointed by him in this behalf together with the audit report thereon shall be forwarded annually to the Central Government.

Annual Report to be laid before parliament

30. The Central Government shall cause the annual report and auditor's report to be laid, as soon as may be after they are received, before each House of Parliament.

CHAPTER - VIII

Finance, Accounts and Audit of State Biodiversity Board

Grants of money by State Government to State Biodiversity Board

31. The State Government may, after due appropriation made by the State Legislature by law in this behalf, pay to the State Biodiversity Board by way of grants or loans such sums of money as the State Government may think fit for being utilized for the purposes of this Act.

Constitution of State Biodiversity Fund

32.(1) There shall be constituted a Fund to be called the State Biodiversity Fund and there shall be credited thereto-

(a) any grants and loans made to the State Biodiversity Board under section 31;

(b) any grants or loans made by the National Biodiversity Authority;

(c) all sums received by the State Biodiversity Board from such other sources as may be decided upon by the State Government.

(2) The State Biodiversity Fund shall be applied for--

(a) the management and conservation of heritage sites;
(b) compensating or rehabilitating any section of the people economically affected by notification under sub-section (1) of section 37;

(c) conservation and promotion of biological resources;

(d) socio-economic development of areas from where such biological resources or knowledge associated thereto has been accessed subject to any order made under section 24, in consultation with the local bodies concerned;

(e) meeting the expenses incurred for the purposes authorized by this Act.

Annual report of State Biodiversity Board

33. The State Biodiversity Board shall prepare, in such form and at such time in each financial year as may be prescribed, its annual report, giving a full account of its activities during the previous financial year, and submit a copy thereof to the State Government.

Audit of accounts of State Biodiversity Board

34. The accounts of the State Biodiversity Board shall be maintained and audited in such manner as may, in consultation with the Accountant-General of the State, be prescribed and the State Biodiversity Board shall furnish, to the State Government, before such date as may be prescribed, its audited copy of accounts together with auditor's report thereon.

Annual Report of State Biodiversity Board to be laid before State Legislature

35. The State Government shall cause the annual report and auditor's report to be laid, as soon as may be after they are received, before the House of State Legislature.

CHAPTER - IX

Duties of the Central and the State Governments

Central Government to develop National strategies plans. Etc., for conservation, etc., of biological diversity

36.(1) The Central Government shall develop national strategies, plans, programmes for the conservation and promotion and sustainable use of biological diversity including measures for identification and monitoring of areas rich in biological resources, promotion of in situ, and ex situ, conservation of biological resources, incentives for research, training and public education to increase awareness with respect to biodiversity.

(2) Where the Central Government has reason to believe that any area rich in biological diversity, biological resources and their habitats is being threatened by overuse, abuse or neglect, it shall issue directives to the concerned State Government to take immediate ameliorative measures, offering such State Government any technical and other assistance that is possible to be provided or needed.

(3) The Central Government shall, as far as practicable wherever it deems appropriate, integrate the conservation, promotion and sustainable use of biological diversity into relevant sectoral or cross-sectoral plans, programmes and policies.

(4) The Central Government shall undertake measures,-

(i) wherever necessary, for assessment of environmental impact of that project which is likely to have adverse effect on biological diversity, with a view to avoid or minimize such effects and where appropriate provide for public participation in such assessment;

(ii) to regulate, manage or control the risks associated with the use and release of living modified organisms resulting from biotechnology likely to have adverse impact on the conservation and sustainable use of biological diversity and human health.

(5) The Central Government shall endeavour to respect and protect the knowledge of local people relating to biological diversity, as recommended by the National Biodiversity Authority through such measures, which may include registration of such knowledge at the local, State or national levels, and other measures for protection, including sui generis system.
Explanation:-For the purposes of this section,–

(a) "ex situ conservation" means the conservation of components of biological diversity outside their natural habitats;

(b) "in situ conservation" means the conservation of ecosystems and natural habitats and the maintenance and recovery of viable populations of species in their natural surroundings and, in the case of domesticated or cultivated species, in the surroundings where they have developed their distinctive properties.

Biodiversity heritage sites

37.(1) Without prejudice to any other law for the time being in force, the State Government may, from time to time in consultation with the local bodies, notify in the Official Gazette, areas of biodiversity importance as biodiversity heritage sites under this Act.

(2) The State Government, in consultation with the Central Government, may frame rules for the management and conservation of all the heritage sites.

(3) The State Government shall frame schemes for compensating or rehabilitating any person or section of people economically affected by such notification.

Power of Central Government to notify threatened species

38. Without prejudice to the provisions of any other law for the time being in force, the Central Government, in consultation with the concerned State Government, may from time o time notify any species which is on the verge of extinction or likely to become extinct in the near future as a threatened species and prohibit or regulate collection thereof for any purpose and take appropriate steps to rehabilitate and preserve those species.

Power of Central Government to designate repositories

39.(1) The Central Government may, in consultation with the National Biodiversity Authority, designate institutions as repositories under this Act for different categories of biological resources.

(2) The repositories shall keep in safe custody the biological material including voucher specimens deposited with them.

(3) Any new taxon discovered by any person shall be notified to the repositories or any institution designated for this purpose and he shall deposit the voucher specimens with such repository or institution.

Power of Central Government to exempt certain biological resources

40. Notwithstanding anything contained in this Act, the Central Government may, in consultation with the National Biodiversity Authority, by notification in the Official Gazette, declare that the provisions of this Act shall not apply to any items, including biological resources normally traded as commodities.

CHAPTER - X

Biodiversity Management Committees

Constitution of Biodiversity Management Committee

41.(1) Every local body shall constitute a Biodiversity Management Committee within its area for the purpose of promoting conservation, sustainable use and documentation of biological diversity including preservation of habitats, conservation of land races, folk varieties and cultivars, domesticated stocks and breeds of animals and microorganisms and chronicling of knowledge relating to biological diversity.

Explanation.- For the purposes of this sub-section,—

(a) "cultivar" means a variety of plant that has originated and persisted under cultivation or was specifically bred for the purpose of cultivation;

(b) "folk variety" means a cultivated variety of plant that was developed, grown and exchanged informally among farmers;

(c) "landrace" means primitive cultivar that was grown by ancient farmers and their successors.

(2) The National Biodiversity Authority and the State Biodiversity Boards shall consult the Biodiversity Management Committees while taking any decision relating to the use of biological resources and knowledge associated with such resources occurring within the territorial jurisdiction of the Biodiversity Management Committee.

(3) The Biodiversity Management Committees may levy charges by way of collection fees from any person for accessing or collecting any biological resource for commercial purposes from areas falling within its territorial jurisdiction.

CHAPTER - XI

Local, Biodiversity Fund

Grants to Local Biodiversity Fund

42. The State Government may, after due appropriation made by State Legislature by law in this behalf, pay to the Local Biodiversity Funds by way of grants or loans such sums of money as the State Government may think fit for being utilized for the purposes of this Act.

Constitution of Local Biodiversity Fund

43.(1) There shall be constituted a Fund to be called the Local Biodiversity Fund at every area notified by the State Government where any institution of self-government is functioning and there shall be credited thereto-

(a) any grants and loans made under section 42;

(b) any grants or loans made by the National Biodiversity Authority;

(c) any grants or loans made by the State Biodiversity Boards;

(d) fees referred to in sub-section (3) of section 41 received by the Biodiversity Management Committees;

(e) all sums received by the Local Biodiversity Fund from such other sources as may be decided upon by the State Government.

Application of Local Biodiversity Fund

44.(1) Subject to the provisions of sub-section (2), the management and the custody of the Local Biodiversity Fund and the purposes for which such Fund shall be applied, be in the manner as may be prescribed by the State Government.

(2) The Fund shall be used for conservation and promotion of biodiversity in the areas falling within the jurisdiction of the concerned local body and for the benefit of the community in so far such use is consistent with conservation of biodiversity.

Annual Report of Biodiversity Management Committees

45. The person holding the custody of the Local Biodiversity Fund shall prepare, in such form and during each financial year at such time as may be prescribed, its annual report, giving a full account of its activities during the previous financial year, and submit a copy thereof to the concerned local body.

Audit of accounts of Biodiversity Management Committees

46. The accounts of the Local Biodiversity Fund shall be maintained and audited in such manner as may, in consultation with the Accountant-General of the State, be prescribed and the person holding the custody of the Local Biodiversity Fund shall furnish, to the concerned local body, before such date as may be prescribed, its audited copy of accounts together with auditor's report thereon.

Annual report, etc, of the Biodiversity Management Committee to be submitted to district Magistrate

47. Every local body constituting a Biodiversity Management Committee under sub-section (1) of section 41, shall cause, the annual report and audited copy of accounts together with auditor's report thereon referred to in sections 45 and 46, respectively and relating to such Committee to be submitted to the District Magistrate having jurisdiction over the area of the local body.

CHAPTER - XII

Miscellaneous

National Biodiversity Authority to be bound by the directions given by Central Government

48.(1) Without prejudice to the foregoing provisions of this Act, the National Biodiversity Authority shall, in the discharge of its functions and duties under this Act, be bound by such directions on questions of policy as the Central Government may give in writing to it from time to time.

Provided that the National Biodiversity Authority shall, as far as practicable, be given opportunity to express its views before any direction is given under this sub-section.

(2) The decision of the Central Government whether a question is one of policy or not shall be final.

Power of State Government to give Directions

49.(1) Without prejudice to the foregoing provisions of this Act, the State Biodiversity Board shall, in the discharge of its functions and duties under this Act, be bound by such directions on questions of policy as the State Government may give in writing to it from time to time:

Provided that the State Biodiversity Board shall, as far as practicable, be given an opportunity to express its views before any direction is given under this sub-section.

(2) The decision of the State Government whether a question is one of policy or not shall be final.

Settlement of disputes between State Biodiversity Boards

50.(1) If a dispute arises between the National Biodiversity Authority and a State Biodiversity Board, the said Authority or the Board, as the case may be, may prefer an appeal to the Central Government within such time as may be prescribed.

(2) Every appeal made under sub-section (1) shall be in such form as may be pre-scribed by the Central Government.

(3) The procedure for disposing of an appeal shall be such as may be prescribed by the Central Government:

Provided that before disposing of an appeal, the parties shall be given a reasonable opportunity of, being heard.

(4) If a dispute arises between the State Biodiversity Boards, the Central Government shall refer the same to the National Biodiversity Authority.

(5) While adjudicating any dispute under sub-section (4), the National Biodiversity Authority shall be guided by the principles of natural justice and shall follow such proce-dure as may be prescribed by the Central Government.

(6) The National Biodiversity Authority shall have, for the purposes of discharging its functions under this section, the same powers as are vested in a civil court under the Code of Civil Procedure, 1908 in respect of the following matters, namely–

(a) summoning and enforcing the attendance of any person and examining him on oath;

(b) requiring the discovery and production of documents;

(c) receiving evidence on affidavits;

(d) issuing commissions for the examination of witnesses or documents;

(e) reviewing its decisions; dismissing an application for default or deciding it ex parte;

(g) setting aside any order of dismissal of any application for default or any order passed by it ex parte;

(h) any other matter which may be prescribed.

(7) Every proceeding before the National Biodiversity Authority shall be deemed to be a judicial proceeding within the meaning of sections 193 and 228, and for the purpose of section 196, of the Indian Penal Code and the National Biodiversity Authority shall be deemed to be a civil court for all the purposes of section 195 and Chapter XXV1 of the Code of Criminal Procedure, 1973.

Members, officers, etc., of National Biodiversity Authority and State Biodiversity Board deemed to be public servants

51. All members, officers and other employees of the National Biodiversity Authority or the State Biodiversity Board shall be deemed, when acting or purporting to act in pursuance of any of the provisions, of this Act, to be public servants within the meaning of section 21 of the Indian Penal Code.

Appeal

52. Any person, aggrieved by any determination of benefit sharing or order of the National Biodiversity Authority or a State Biodiversity Board under this Act, may file an appeal to t ' he High Court within thirty days from the date of communication to him, of the determination or order of the National Biodiversity Authority or the State Biodiversity Board, as the case may be:

Provided that the High Court may, if it is satisfied that the appellant was prevented by sufficient cause from filing the appeal-within the said period, allow it to be filed within a further period not exceeding sixty days.

Execution of determination or order

53. Every determination of benefit sharing or order made by the National Biodiversity Authority or a State Biodiversity Board under this Act or the order made by the High Court in any appeal against any determination or order of the National Biodiversity Authority or a State Biodiversity Board shall, on a certificate issued by any officer of the National Biodiversity Authority or a State Biodiversity Board or the Registrar of the High Court, as the case may be, be deemed to be decree of the civil court and shall be executable in the same manner as a decree of that court.

Explanation.- For the purposes of this section and section 52, the expression "State Biodiversity Board" includes the person or group of persons to whom the powers or functions under sub-section (2) of section-22 have been delegated under the proviso to that sub--section and the certificate relating to such person or group of persons under this section shall be issued by such person or group of persons, as the case may be.

Protection of action taken in good faith

54. No suit, prosecution or other legal proceedings shall lie against the Central Government or the State Government or any officer of the Central Government or the State Government or any member, officer or employee of the National Biodiversity Authority or the State Biodiversity Board for anything which is in good faith done or intended to he done under this Act or the rules or regulations made thereunder.

Penalties

55.(1) Whoever contravenes or to or abets the contravention of the provisions of section 3 or section 4 or section 6 shall be punishable with imprisonment for a term which may extend to five years, or with fine which may extend to ten lakh rupees and where the damage caused exceeds tend lakh rupees such fine may commensurate with the damage caused, or with both.

(2) Whoever contravenes or attempts to contravene or abets the contravention of the provisions of section 7 or any order made under sub-section (2) of section 24 shall be punishable with imprisonment for a term which may extend to three years, or with fine which may extend to five lakh rupees, or with both.

Penalty for contravention of directions or orders of Central government, State government, National Biodiversity Authority and State Biodiversity Boards

56. If any person contravenes any direction given or order made by the Central Government, the State Government, the National Biodiversity Authority or the State Biodiversity Board for which no punishment has

been separately provided under this Act, he shall be punished with a fine which may extend to one lakh rupees and in case of a second or subsequent offence, with fine which may extend to two lakh rupees and in the case of continuous contravention with additional fine which may extend to two lakh rupees everyday during which the default continues.

Offences by companies

57.(1) Where an offence or contravention under this Act has been committed by a company, every person who at the time the offence or contravention was committed was in charge of, and was responsible to, the company for the conduct of the business of the company, as well as the company, shall be deemed to be guilty of the offence or contraven-tion and shall be liable to be proceeded against and punished accordingly:

Provided that nothing contained in this sub-section shall render any such person liable to any punishment provided in this Act, if he proves that the offence or contravention was committed without his knowledge or that he had exercised all due diligence to prevent the commission of such offence or contravention.

(2) Notwithstanding anything contained in sub-section (1), where an offence or con-travention under this Act has been committed by a company and it is proved that the offence or contravention has been committed with the consent or connivance of, or is attributable to, any neglect on the part of any director, manager, secretary or other officer of the company, such director, manager, secretary or other officer shall also be deemed to be guilty of the offence or contravention and shall be liable to be proceeded against and punished accordingly.

Explanation. -For the purposes of this section,—

(a) "company" means any body corporate and includes a firm or other associa-tion of individuals; and

(b) "director", in relation to a firm, means a partner in the firm.

Offences to be cognizable and non-bailable

58. The offences under this Act shall be cognizable and non-bailable.

Act to have effect in addition to other Acts

59. The provisions of this Act shall be in addition to, and not in derogation of, the provisions in any other law, for the time being in force, relating to forests or wildlife.

Power of Central Government to giver directions to State Government

60. The Central Government may give directions to any State Government as to the carrying into execution in the State of any of the provisions of this Act or of any rule or regulation or order made thereunder.

Cognizance of offences

61. No Court shall take cognizance of any offence under this Act except on a com-plaint made by—

(a) the Central Government or any authority or officer authorized in this behalf by that Government; or

(b) any benefit claimer who has given notice of not less than thirty days in the prescribed manner, of such offence and of his intention to make a complaint, to the Central Government or the authority or officer authorized as aforesaid.

Power of Central Government to make rules

62.(1) The Central Government may' by notification in the Official Gazette, make rules for carrying out the purposes of this Act.

(2) In particular, and without prejudice to the generality of the foregoing power, such rules may provide for A or any of the following matters, namely:–

(a) terms and conditions of service of the Chairperson and members under section 9;

(b) powers and duties of the Chairperson under section 10;

(c) procedure under sub-section (1) of section 12 in regard to transaction of business at meetings,

(d) form of application and payment of fees for undertaking certain activities under sub-section (1) of section 19;

(e) the form and manner of making an application under sub-section (2) of section 19;

(f) form of application and the manner for transfer of biological resource or knowledge under sub-section (2) of section 20;

(g) form in which, and the time of each financial year at which, the annual report of the National Biodiversity Authority shall be prepared and the date before which its audited copy of accounts together with auditor's report thereon shall be furnished under section 28;

(h) form in which the annual statement of account shall be prepared under subsection (1) of section 29;

(i) the time within which and the form in which, an appeal may be preferred, the procedure for disposing of an appeal and the procedure for adjudication, under section 50;

(j) the additional matter in which the National Biodiversity Authority may exercise powers of the civil court under clause (h) of sub-section (6) of section 50;

(k) the manner of giving notice under clause (b) of section 61;

(1) any other matter which is to be, or may be, prescribed, or in respect of which provision is to be made, by rules.

(3) Every rule made under this section and every regulation made under this Act shall be laid, as soon as may he after it is made, before each House of Parliament, while it is in session or a total period of thirty days which may be comprised in one session or in two or more successive sessions, and if, before the expiry of the session immediately following. The session

or the successive sessions aforesaid, both Houses agree in making any modification in the rule or regulation or both Houses agree that the rule or regulation should not be made, the rule or regulation shall thereafter have effect only in such modified form or be of no effect, as the case may be; so, however, that any such modification or annulment shall be without prejudice to the validity of anything previously done under that rule or regulation.

Power of State Government to make rules

63.(1) The State Government may, by notification in the Official Gazette, make rules for carrying out the purposes of this Act.

(2) In particular, and without prejudice to the generality of the foregoing power, such rules may provide for all or any of the following matters, namely:–

(a) the other functions to be performed by the State Biodiversity Board under clause (c) of section 23;

(b) the form in which the prior intimation shall be given under sub-section (1) of section 24;

(c) the form in which, and the time of each financial year at which, the annual report shall be prepared under section 33;

(d) the manner of maintaining and auditing the accounts of the State Biodiversity Board and the date before which its audited copy of the accounts together with auditor's report thereon shall be furnished under section 34;

(e) management and conservation of national heritage sites under section 37;

(f) the manner of management and custody of the Local Biodiversity Fund and the purposes for which such Fund shall be applied under sub-section (1) of section 44;

(g) the form of annual report and the time at which such report shall be prepared during each financial year under section 45;

(h) the manner of maintaining and auditing the accounts of the Local Biodiversity Fund and the date before which its audited copy of the accounts together with auditor's report thereon shall be furnished under section 46;

(i) any other matter which is to be, or may be, specified.

(3) Every rule made by the State Government under this section shall he laid, as soon as may be after it is made, before each House of the State Legislature where it consists of two Houses, or where such Legislature consists of one House, before that House.

Power tof remove difficulties

64. The National Biodiversity Authority shall, with the previous approval of the Central Government, by notification in the Official Gazette, make regulations for carrying out the purposes of this Act.

Power to remove difficulties

65.(1) If any difficulty arises in giving effect to the provisions of this Act, the Central Government may, by order, not inconsistent with the provisions of this Act, remove the difficulty:

Provided that no such order shall be made after the expiry of a period of two years from the commencement of this Act.

(2) Every order made under this section shall be laid, as soon as may be, after it is made, before each House of Parliament.

Glossary of Important Terms

Allele: Any of the different forms of a gene which may occupy the same locus on a chromosome.

Alpha diversity: The diversity within a particular area or ecosystem, and is usually expressed by the number of species (i.e. species richness) in that ecosystem.

Area of occupancy: Area within its extent of occurrence occupied by a taxon. Size of such area is as of regenerating/colonial/migratory/nesting sites.

Benefit claimers : The conservers of biological resources and holders of knowledge and information from grassroot innovations to practices.

Beta diversity: The change in species diversity between the ecosystem.

Biodiversity Heritage Sites: Areas of biodiversity importance notified by the government.

Biodiversity Management Committee: The Committee constituted by a local body (panchyat, municipalities)within its area for the purpose of promoting conservation, sustainable use and documentation of biological diversity including preservation of habitats, conservation of land races, folk varieties and cultivars, domesticated stocks and breeds of animals and microorganisms and chronicling of knowledge relating to biological diversity.

Biodiversity Register: Meant to document the knowledge of occurrence, practices of propagation, sustainable utilization and conservation, as well as economic uses of biodiversity and genetic resources typical of India.

Biogeography: The scientific study of the geographic distribution of organisms.

Biological diversity or Biodiversity: The variability among living organisms from all sources and the ecological complexes of which they are part, and includes diversity within species or between species and of their habitats.

Biological resources or Bioresources: Plants, animals and micro-organisms or parts thereof, their genetic material and by-products with actual or potential use or value.

Biosphere reserves: These are undisturbed natural areas suitable for scientific study and set aside for research and habitat preservation. Established under UNESCO's Man and the Biosphere (MAB) Programme, where biosphere reserves are a series of protected areas linked through a global network, intended to demonstrate the relationship between conservation and development.

Biosphere: Includes soil, water, and the atmosphere.

Biotechnology: Technological application that uses biological systems for specific use.

Chromosome: Thread like body found within the nucleus with DNA (deoxyribonucleic acid) and protein sheath, and the genes.

Climax: A terminal stage of ecological succession.

Commercial utilization: Bio-resources for commercial utilization.

Community: Life forms in close interdependent and common environment.

Conservation Dependent : Lower risk IUCN category on threats under species-specific or habitat-specific conservation program.

Conservation values: The value to society of conserving environmental resources.

Conservation: Use and management of biological resources for the welfare of the society and posterity.

Cost-benefit analysis : The appraisal of an investment project which includes all social and financial costs and benefits accruing to the project.

Country of origin of genetic resources: The country which possesses those genetic resources in *in-situ* conditions.

Country providing genetic resources: The country supplying genetic resources collected from *in-situ* sources, including populations of both wild and domesticated species, or taken from *ex-situ* sources, which may or may not have originated in that country.

Critically Endangered : Facing an extremely high risk of extinction in the wild in the immediate future.

Cry-preservation: Maintaining tissues or seeds for the purpose of long term storage at subzero temperature or in liquid nitrogen.

Cultivar: Variety of plant that has originated through cultivation.

Data Deficient : With inadequate information to make a direct or indirect assessment of its risk of extinction.

Domestic biodiversity: The genetic variation existing among the species, breeds and cultivars under domestication.

Domesticated or cultivated species: The species in which the evolutionary process has been influenced by man to meet his needs.

Domestication: The process by which plants, animals or microbes selected from the wild by man for domestication.

Ecosystem diversity: Existence of different land use and habitat variability in a ecosystem.

Ecosystem: Dynamic complex of flora and fauna and inanimate components interacting with them as a functional unit.

Ecotone: Transition zone between the habitats or distribution ranges of species.

Endangered : In danger of extinction and for which survival is unlikely if the causal factors of loss continue. Not critically endangered but facing a very high risk of extinction in the wild in the near future.

Environmental impact assessment: Process of evaluating the likely environmental impacts of a proposed development, taking into account inter-related socioeconomic, cultural and human health impacts, both beneficial and adverse.

Ex situ conservation: Removal of individual species from its site of natural occurrence, i.e. conservation off site.

Exotic: A species in an area in which it does not naturally occur.

Extinct : There is no reasonable doubt that the last individual has died.

Extinct in the wild: Extinct in the wild. No more found in the wild.

Extent of occurrence: The area contained within the shortest continuous imaginary boundary, which can be drawn to encircle all the known sites of present occurrence of a taxon excluding the 'escape' from wilderness. Measured by a minimum convex polygon containing all the sites of occurrence.

Flagship or umbrella species: Popular, charismatic species serving as symbols and rallying points to stimulate conservation awareness and action.

Fragmentation: Extensive landscape features broken into disjunct, isolated, patches as a result of land-use changes and patterns.

Gamma diversity: A measure of the overall diversity for the different ecosystem within a region.

Gene Bank: A storage facility where germplasm is stored in the form of seeds, pollen, embryos, semen, pollen, or *in vitro* culture, or in cryogenic storage, or, in the case of a field gene bank, as plants growing in the field.

Gene Flow: The movement of genes between populations as the result of out crossing and natural selection.

Gene Pool: The total sum of genetic material of an interbreeding population.

Gene: The basic functional unit of inheritance.

Genetic diversity: The species or population in possession of a variety of genetic traits.

Genetic erosion: Loss of genetic diversity between and within populations due to human intervention, environmental changes etc.

Genetic resources: Genetic material of real or potential value.

Genome: All the genes of a particular organism or species.

Genotype: Genetic constitution of an organism.

Germplasm: Hereditary material transmitted to offspring through the germ cells. A collection of many different varieties and species representing a diverse collection of genotypes.

Green book: It enlists rare plants growing in protected areas like Botanic Gardens.

Habitat: The place or type of site where an organism or population naturally occurs.

Heterozygous: Having one or more unlike alleles at corresponding loci of homologous chromosomes.

Homozygous: Having like alleles for a particular gene at corresponding loci on homologous chromosomes.

Hotspots: Regions with a great diversity of endemic species and significantly impacted.

Hybridization: The process of crossing individuals that possess different genctic makeup.

In situ conservation: Conservation of ecosystems and natural habitats. Refers to the conservation of genetic resources of target species, on site, within the natural or original ecosystem in which they occur, or on the site formerly occupied by that ecosystem.

Indeterminate : Insufficient information presently available with respect to any of Rare, Endangered and categories.

Indigenous communities : Tribal people or original inhabitants of a place or living in conformity with their own social, economic, and cultural customs and traditions.

Intellectual property right: A right enabling an inventor to exclude imitators from the market for a stipulated period.

Introduced species: A species occurring in an area outside its natural range. Also known as an exotic or alien species.

Invasive species: An introduced species which invades natural and modified habitats and landscapes.

Inventorization or inventorying: Surveying, sorting, cataloguing, quantifying and mapping of entities such as individuals, populations, species, habitats, ecosystems and landscapes or their components for analysis of process and consolidation of information.

Keystone species: A species whose loss from an ecosystem would cause a greater than average change in other species populations or ecosystems processes.

Landrace : Primitive cultivar that was grown by ancient farmers.

Landscape: The mosaic of various land uses, cover in a specific area.

Least Concern : Not qualifying for conservation dependent or near threatened category.

Local Bodies: Panchayats and Municipalities, by whatever name called, within the meaning of clause (1) of article 243B and clause (1) of article 243Q of the Constitution of India and in the absence of any Panchayats or Municipalities, institutions of self-government constituted under any other provision of the Constitution or any Central Act or State Act. In context to National Biodiversity Authority, the local body can constitute Biodiversity Management Committees.

Lower Risk : A taxon is lower risk when it has been evaluated and does not satisfy the criteria for any of the categories critically endangered, endangered, or vulnerable.

Mature individuals: Severely fragmented (isolated sub population) refers to the pronounced extinction risks with a reduced probability of regeneration.

Micro propagation: The clonal production of trees or other plants through techniques of in vitro culture of buds, plantlets, or tissues.

Mutation: Sudden heritable change in the gene or chromosome constitution causing change in number, structure, size or sequence.

National Biodiversity Authority: The National Biodiversity Authority established under section 8 of the Biodiversity Act- 2002 of India.

National Parks: A relatively large area of land set aside (under some Act) for its features of predominantly unspoiled natural landscape, flora and fauna, and permanently dedicated for public enjoyment, education and inspiration and protected from all interference.

Near Threatened : Taxa not qualifying for conservation dependent but which closer to vulnerable category.

Neo-endemics: Species and subspecies that have evolved relatively recently.

Not Evaluated : Not yet assessed against any of the categorizing criteria.

Out of danger : Formerly included in one of the threat categories, but now considered relatively secure.

Phenotype: Observable character of an individual resulting from interaction of the genotype with the environment.

Phylogenetic: Evolutionary history of a particular group of organisms.

Population: A group of interbreeding individuals occupying a particular area. The total number of mature individuals of a taxon.

Porosity: Measure of number of patches or density of patches within a specific kind.

Protected area: A geographically defined area, which is designated or regulated and managed to achieve specific conservation objectives.

Provenance: Geographical entity of a species from an identifiable region in the natural habitat of a species.

Rare: Elements usually localized within restricted geographical areas or habitats or are thinly scattered over a more extensive range.

Red Data Book: Book dealing with threatened plants or animals of any region.

Reproduction Constraints: Taxon with diagnostic features on the reproductive parts influencing the survival.

Seed Bank: A facility designed for the *ex situ* conservation of plant samples through seed preservation and storage.

Species diversity: The number and variety of species found in a given area in a region.

Species richness: The number of species within a region. A term commonly used as a measure of species diversity, but technically only one aspect of diversity. Refers to alpha diversity.

Species: A taxonomic subdivision of the genus. A group of organisms that actually or potentially interbreeds and reproductively isolated from other such group.

State Biodiversity Board: Stands for the State Biodiversity Board established under section 22 of the Biodiversity Act 2002.

Sub-population: Geographical or otherwise distinct groups.

Sustainable development: Development that meets the needs and aspirations of the current generation without compromising the ability to meet those of future generations.

Sustainable management: The term can only be precisely defined in terms of the management objectives of a particular forest, however, it should always incorporate the principles of sustained development, to meet the needs of the present without compromising the ability of future generations to meet their own needs. It has essential objectives in common with the conservation of genetic resources.

Taxon (pl. taxa) **:** A classificatory unit to which individuals, or set of species, are assigned.

Tetraploid: Referring to a cell or organism containing four sets of chromosomes.

Threatened Taxa: Species for which the potential for loss exists. Also refers to a species in one of the three categories, namely Critically endangered, Endangered and Vulnerable.

Tissue culture: A technique for cultivating cells, tissues or organs of plants in a sterile, synthetic medium (also '*in vitro* culture').

Total Economic Value: The sum of use and non-use values with due consideration of any trade-offs or mutually exclusive uses or functions of the resource/habitat in question.

Traditional Knowledge: Refers to the traditional knowledge, innovations and practices of indigenous and local communities embodying traditional lifestyles relevant for the conservation and sustainable use of biological diversity.

Valuation: Method for determining the importance of environmental consequences of economic activity.

Value added products: The products that may contain portions or extracts of plants and animals in unrecognizable and physically inseparable form.

Vegetation: A striking expression of various collective interdependent and coexisting plant communities (flora) in a landscape.

Vulnerable: Not critically endangered or endangered but facing a high risk of extinction in the wild in the medium-term future.

Wild relative: Plant or animal species that are taxonomically related to crop or livestock species and serve as potential sources for genes in breeding of new varieties of those crops or livestock.

Wildlife sanctuary: Similar to a National Park but dedicated to protect the wildlife, and onsiders the conservation of species only.

Abbreviations

AICPE	:	All India Co-ordinated Project on Ethnobiology
AK	:	Advance Knowledge
APHIS	:	Animal and Health Inspection Service of USA.
BC	:	Biodiversity Conservation
BG	:	Botanical Garden
BR	:	Biosphere Reserve
CAMP	:	Conservation Assessment and Management Plan
CBBIA	:	Capacity Building and Impact Assessment
CBD	:	Convention on Biological Diversity
CBSG	:	Conservation Breeding Specialist Group
CCAMLR	:	Conference on the Conservation of Antarctic Marine Living Resources.
CDS	:	Conservation Dependent Species
CGD	:	Centre of Genetic Diversity
CSD	:	Centre of Species Diversity
CIS	:	Commonwealth of Independent States (erstwhile USSR)
CITES	:	Convention on International Trade in Endangered Species of wild fauna and flora
COP	:	Conference of Parties
CR	:	Critically Endangered
CRH	:	Critically Endangered Habitat
DD	:	Data Deficient
DDH	:	Data Deficient Habitat
DNA	:	Deoxyribose Nucleic Acid
E	:	Extinct
EEC	:	European Economic Community
EH	:	Endangered Habitat
EIA	:	Environmental Impact Assessment

EN	:	Endangered
ER	:	Environmental Reports
EUFORGEN	:	Noble Hardwoods Network of European Forest Genetic Resources Programme
EW	:	Extinct in Wild
EX	:	Extinct
FAO	:	Food and Agriculture Organization
GIS	:	Geographic Informaation System
GMOs	:	Genetically Modified Organisms
GPS	:	Global Positioning System
I	:	Indeterminate
IA	:	Strengthening participatory will on the integration of indigenous and advance knowledge holders
IBS	:	Intermediate Biotechnology Service
IC	:	Involvement of indigenous communities
ICGEB	:	International Centre for Genetic Engineering and Biotechnology
IK	:	Indigenous Knowledge
IKH	:	Indigenous Knowledge Holders
IPRs	:	Intellectual Property Rights
ISABA	:	International Service for the Acquisition of Biotech Applications.
ISNAR	:	International Service for National A griculture Research
ITK	:	Indigenous Traditional Knowledge
IUCN	:	International Union for Conservation of Nature and Natural Resources
JFM	:	Joint Forest Management
KAC	:	Key Areas of Conservation
KNTK	:	Karnataka
KRL	:	Kerala

LC	:	Least Concern
LCH	:	Least Concern Habitat
LR	:	Lower Risk
MMT	:	Million Metric Tons
MPAS	:	Marine Protecsted Areas
NAS	:	National Academy of Sciences, Washington
NBRI	:	National Botanical Research Institute, Lucknow
NE	:	Not Evaluated
NEPA	:	US National Environmental Policy Act
NP	:	National Park
NPP	:	National Population Policy
NT	:	Near Threatened
O	:	Taxon out of Danger
OH	:	Habitat out of Danger
PCR	:	Polymerase – Chain – Reaction
PHVA	:	Population and Habitat Viability Assessment
POSSCEF	:	Project on Study, Survey and Conservation of Endangered species of Flora
R & D	:	Research and Development
R	:	Rare
RCH	:	Restoration or Recharging Constrained Habitat
RDH	:	Restoration Dependent Habitat
RFLP	:	Restriction - Fragment – Length – Polymorphisms
REP	:	Reproduction Constraint
RH	:	Rare Habitat
SAARC	:	South Asia Association for Regional Cooperation
SSC	:	Survival Service Commission
T	:	Threatened
TH	:	Threatened Habitat
TN	:	Tamil Nadu

TPC	:	Threatened Plants Committee
TRAFFIC	:	Trade Related Analysis on Flora and Fauna in Commerce
UNCEIA	:	United Nations Convention on Environmental Impact Assessment
UNEP	:	United Nations Environmental Programme
UNESCO	:	United Nations Educational, Scientific and Culture Organization
UNIDO	:	United Nations Industrial Development Organization
UNCLS	:	United Convention on the Law of the Sea
USAID	:	United States Agency for International Development
UTs	:	Union Territories
VH	:	Vulnerable Habitat
VU	:	Vulnerable
WCMC	:	World Conservation Monitoring Centre
WHO	:	World Health Organization
WII	:	Wildlife Institute of India, Dehra Dun
WS	:	Wildlife Sanctuary
WTO	:	World Trade Organization

INDEX